Against Instinct

Against Instinct

From Biology to Philosophical Psychology

Dennis M. Senchuk

Temple University Press
Philadelphia

Temple University Press, Philadelphia 19122
Copyright © 1991 by Temple University. All rights reserved
Published 1991
Printed in the United States of America

The paper used in this publication meets the minimum
requirements of American National Standard for Information
Sciences—Permanence of Paper for Printed Library Materials,
ANSI Z39.48-1984 ∞

Library of Congress Cataloging-in-Publication Data
Senchuk, Dennis M.
Against instinct : from biology to philosophical psychology /
Dennis M. Senchuk.
p.   cm.
Includes bibliographical references and index.
ISBN 0-87722-815-9 (alk. paper)
1. Instinct.   2. Behavior genetics.   3. Instinct (Philosophy)
4. Free will and determinism.   I. Title.
BF685.S46   1991
128′.3—dc20   90-47638

To Karen Hanson and our darling
daughters, Chloë Miranda and
Tia Elizabeth

A shipwrecked man cast up on the beach fell asleep after his struggle with the waves. When he woke up, he bitterly reproached the Sea for its treachery in enticing men with its smooth and smiling surface, and then, when they were well embarked, turning in fury upon them and sending both ship and sailors to destruction. The Sea arose in the form of a woman, and replied, "Lay not the blame on me, O sailor, but on the Winds. By nature I am as calm and safe as the land itself: but the Winds fall upon me with their gusts and gales, and lash me into a fury that is not natural to me."

—*Aesop*

Contents

Preface

Despite many venerable philosophical and scientific efforts at self-understanding, we earthly agents move in still mysterious ways our behavioral wonders to perform. We may estimate that we act at least partly of our own volition, but the exact measure of our action and volition has not been taken. It may always be somehow in our natures to behave as we do, but whether we are like Aesop's Sea, largely goaded by prevailing winds to do the specific things we do, or whether we are already predetermined, biologically, to do them remains obscure.

Those who want full credit for their acts sometimes assert, even so, that their acts obey their instincts. Such talk, after Darwin, might seem to imply that the credit properly belongs to evolutionary processes, not people; but these processes may be interpreted as a (nontranscendent) kind of providence and one's instincts as one's most personal natural tendencies. Even the apparent opposition between instinct and conscious design need not erode the curious pride one takes in one's allegedly instinctive acts, for they may still be seen (not as blindly brutish productions) as less superficially contrived, truer expressions of one's essence than are the results of reason.

Instead of being thought at odds with our distinctively human nature, instincts are often held to constitute it. And though they might appear merely to sanction the social status quo, our common native tendencies are sometimes said to be our only sure defense against tyrannical manipulations of our bodies politic. Much, then, can be said to recommend instincts, if we are of a mind to do so. Should we be? Are there any instincts in the first place?

Contemporary behavioral and cognitive sciences do show significant theoretical allegiance to the notion of innate behavior. What seems to enjoy a vogue is talk of instincts (especially social ones) and innate structures, the latter alleged to be the "hard-wired" neurological basis for linguistic and (perhaps other) intellectual functioning. And it would be a mistake to dismiss these trends as mere fashion, of little substance—if only because the contemporary discussion is closely tied to what is arguably the deepest and most enduring preoccupation of philosophy: the controversy between rationalism and empiricism.

Were we to trace the origins of this debate, we should have to mark Plato's *Meno* as its first full surviving introduction. Were we to look back to a less ancient high point, to the epistemologically oriented systems of the post-Cartesian period of philosophy, we should find that same debate as a principal focus of the writings of Leibniz and Locke, Kant and Hume, and other grand contestants in the quarrelsome course of philosophizing about human nature and understanding.

Yet a preoccupation is not always a permanent occupation of intellectual territory, and the great epistemological systems have been viewed in recent times, during the heyday of "philosophical analysis," as ill-conceived fusions of philosophical and psychological parts. It was assumed that a final resolution of the rationalism-empiricism debate would be the corollary of eventual victories in the direct competition among fully scientific theories; and philosophers, once champions for the opposing sides in the debate, were strangely content to serve as mere referees, rendering only preliminary decisions about the ongoing contest.

Times change, though, and the opinion is now widespread that there is no clear distinction between philosophy and science, between conceptual and empirical investigations. Philosophers, seeing an opportunity, are hastening to "naturalize" their inquiries, to reconcile their claims with those of science. Too often, this amounts to little more than the adoption of a quasi-scientific idiom, in which the term *cause* figures far more prominently than it does in scientific writings. Too seldom is science taken seriously enough, where seriousness is gauged by the degree of critical scrutiny accorded to the putative findings of scientists. One suspects a less than complete conversion to the idea that science and philosophy are no more distinct than Aristotle, for one, assumed them to be—so, for example, the complaint "That is an empirical question!" has hardly been banished from philosophical dialogue. But perhaps the real problem is an almost Lear-like relinquishment of final (epistemological and metaphysical) authority.

Philosophy may no longer claim hegemony over its descendant sciences, but there is no good reason for philosophers not to reenter the fray in ongoing contests of ancestral concern to them. In this book, I mean to take sides against nativism, especially though not exclusively with respect to behavior. I want to subdue, to the fullest extent possible, the speculative impetus in favor of instinct positing. Since the battle has been waged in this century most vigorously in science, in what might most comprehensively be called "behavioral biology," I will turn first to that field.

My overall campaign is a skeptical critique. As a late arrival on the field of battle, I confront a daunting array of experimental findings and other theoretical considerations fortifying the instinctivity hypothesis. Yet that array does afford some conveniently visible targets; and there is besides them an impressive group of redoubts—some now abandoned—that can be reconstructed, linked together to form an effective contravallation.

I should issue the philosophically customary disclaimer that I am not trying to give a historically sound account of the debate between nativists and antinativists, and I must admit that the writers I have selected for consideration might well not be those that would be selected by each side to represent its views. Nonetheless, I do consider the range of views I select to represent fairly the range of available existing options, and I am confident about the correctness of my portrayals of the views themselves.

Still, the debate I set out is something of a fiction, since the assorted scientific and philosophic protagonists I play off against one another were not all participants in an actual exchange of arguments and opinions. As a philosopher, I have been particularly intent upon covering all the "logical space," intent upon considering all the possible methodological and theoretical moves in support of or antagonistic to the positing of instincts. So, for example, though I have selected some ethological theorists—Lorenz, Tinbergen, and others—who are indisputably central to the innateness controversy, I have not limited myself to figures who have been influential for ethology. Koffka, Köhler, and Dewey are considered by me because of their special logical relevance to the biologically, psychologically, and philosophically oriented moves I wish to make against instinct.

Practicing behavioral biologists might be inclined to regard my efforts as still excessively historical and so of little current interest to them. For instead of attending to the very latest scientific accounts of animal behavior, many of which are not so recognizably nativistic or antinativistic, I have sought the locus classicus of each position I consider. Given my purposes, the conceptual and methodological explicitness of the writings—as well as the clarity of their stands on the innateness controversy—is far more important than how recent they happen to be. This feature of my work is by no means untypical of critically oriented writings in the philosophy of science. (Compare, e.g., philosophy of physics, which still profitably attends to the merits of the Copenhagen interpretation of quantum mechanics, despite over half a century of rich experimental and theoretical developments in the field.)

Such considerations, though, may not prove compelling to a more straightforwardly scientific audience; many scientists (including quantum physicists), not to mention philosophers of science, are not persuaded that there is much scientific utility in philosophizing about science. But behavioral biologists are mistaken if they believe that nativism is no longer an issue for them to worry about, that they have gone beyond all that in their more sophisticated treatments of their subject matter. Instinct, my black beast, is less a dead horse than a phoenix; and lest anyone think, say, that an epigenetic view of the ontogenesis of behavior has dealt a manifestly fatal blow to the pretensions of nativism, I begin Part One with a quote from P. B. and J. S. Medawar that should dispel this thought.

Notwithstanding my critical emphasis on conceptually and methodologically explicit writings, I am not offering an old-style analytic (conceptual-methodological) critique of instinct positing. Rather, I am presenting a variant of a still older philosophical maneuver, a Pyrrhonian skeptical critique. This skeptical strategy, which I associate most closely (in the modern era of philosophy) with David Hume's approach in *Dialogues Concerning Natural Religion*, will involve balancing against that nativistic positing some equal and opposite antinativistic hypotheses—equally consonant with the range of evidence alleged to support the idea of instinct but thoroughly opposed to that idea.

Insofar as even the simplest observational and experimental data are construed as theory-laden conjectures, perhaps it is only natural to suppose that it takes one scientific theory to oppose another. Part One of this book proceeds largely as if on that supposition, battering away at pro-instinctive theories from the vantage point of anti-instinctive ones. Pending further empirical reinforcement, some of those latter (interactionist-interpenetrationist) theories do seem fated to withstand the "slings and arrows" of the former and, finally, to prevail; but I am not disposed to declare a victory in the larger contest, since I suspect these theories of harboring some insidious nativism of their own.

Among the more prominent features of Part One are an attack on Lorenzian deprivation experiments as a way of trying to identify innate behaviors; a critique of the ethological reliance on information theory as a way of understanding evolutionary issues about behavior; some critical remarks on individuating behaviors; a (preliminary) discussion of mechanism, prehension, and intelligent functioning; a critical comparison of the Gestalt and Deweyan theories of intelligence; and my sympathetic reappraisal of Zing-Yang Kuo's somewhat neglected theory of behavioral epigenesis.

With the end of Part One, the discussion might be thought to get more philosophically than scientifically interesting; and some scientists might be inclined to stop reading at this point, just where some philosophers might prefer to start. Still, the skepticism in both parts is addressed to both sets of readers. Philosophers too frequently question everything but scientific "fact," even as they deny a real distinction between nonempirical and empirical truths; and scientists have too often been unconcerned about the conceptual flimsiness of their own architecture, its vulnerability to argumentative onslaught—as if they truly believed those philosophical sycophants who assure them that someday everyone will adopt their principles of construction. But as I hope to suggest in Part One, even ground-level experimentally confirmed facts about the biology of behavior may not be what they seem; and, as I come to insist in Part Two, there are some significantly prior philosophical constraints on scientific conceptualizations of animal behavior.

Both parts of the book testify to the manifold interconnectedness of scientific and philosophical issues, but in Part Two I do regroup on a less evidently scientific basis, purging my allied ranks of as many enemy sympathizers as possible, then constructing and defending a more philosophically motivated counterwork against instinct.

Part Two begins with an effort to elucidate the conceptual framework of the strongest available biotheoretical case against instinct. I first clarify and disambiguate some dialectically crucial notions of (deterministic) plasticity. Then, following Braithwaite's lead but diverging from his developed account, I use these notions to explicate nonpurposive teleology. I find, though, that not only is there a conceptual compatibility between instinct and teleological plasticity, but that (as I argue somewhat later) the notion of an instinct can be plausibly recharacterized, relative to what I call a "teleological automaton," in a way that might survive the objections of some of its severest biological critics.

I explore in some detail the possibility that the notion of intelligence might afford further opposition to instinctivity. Arguing against a purely dispositional view of intelligent behaviors, I nevertheless use Ryle's account of them to frame my alternative suggestion about the indispensable role of consciousness—specifically, conscious readiness—in the production of all truly purposive behaviors.

This suggestion marks a turn to more constructive phases of the inquiry. Skepticism, one of the purest forms of philosophical thinking, is hardly the barren, chaste destroyer it is often reputed to be. Who, though, is the skeptic's most plausibly congenial member of the Pantheon? Surely not Athena, protector of citadels—the one tradition usually assigns to philosophers. Perhaps the best candidate is Artemis, patroness of hunters. But consider her checkered vita: Ever intrigued by young and wild living things, she starts out as a mother-deity, subsequently assists others in giving birth, and late in life acquires an incongruous distinction in the Pantheon as a virginal huntress, goddess of the loud chase. Even Olympian reputations, it seems, should not always be accepted at face value.

My own initial efforts in this book might earn me, figuratively speaking, a bad reputation as a clamorous hound of Artemis: Exploiting any available scientific or philosophical resources, my main strategy might appear, early on at least, to be one of sheer brutish tenacity. But the later phases of my inquiry cannot so easily be depicted as a simple, wild hunt. An impressive fortress protects the nativists, and I must do substantial building and fortifying of my own if I am to have any hope of vanquishing them.

What I contrive for my argumentative purposes is a positive philosophy of mind and action—a decidedly nonnativistic philosophical psychology: Consciousness is characterized in terms of active prehensiveness, and action is accounted for as conscious readiness in furtherance of a plan (of sorts), a "teleological scheme." This consciousness is said to confer a certain flexibility on all purposive behaviors, and that flexibility is the key to ultimate victory over instinct.

In taking sides against instinct, against nature, I might seem forced, in consequence of the usual dichotomy, to come down on the side of nurture, instead. But in my rethinking of the nature-nurture debate, I try to reorient it along a different axis, by constructing an alternative to both options. My alternative is a least instinctive hypothesis that rests upon the notion of behavioral "flexibility." The terminology is by no means new, but I believe my account of this notion and my defense of it are.

While Part One goes against the tide of some well-regarded biobehavioral, principally scientific, theorizing, Part Two takes arms against a sea of philosophical troubles as well. Teleological automata could, I suggest, do a tolerably decent job of mimicking purposive behaviors; and, under certain conditions of ontogenesis, some of these mimetic behaviors could qualify as innate. But are people and other animals thus automatistic? I suggest that they are not; yet rendering that suggestion cogent takes some doing. It requires, inter alia, taking issue with the dogma of determinism and the dogma of its supposed compatibility with free will. The whole enterprise is, if nothing else, a fair indication of how very far one has to go, dialectically, to combat instinct, its kith, and its kin.

Now if philosophers can ill afford to disregard scientific developments, it would seem equally foolish to ignore philosophy's progressive refinements of its own tools. So, even though the issues about determinism are perennial, my discussion of them takes advantage of some recent conceptual developments in philosophy. In particular, I exploit the notion of psychophysical supervenience—though instead of using it (as J. Kim does) to argue in favor of microlevel on up physical determinism, I use it to urge against that view. My defense of higher-order indeterminacy also makes use of some older, Russellian conceptual resources—specifically, "neutral monism" and the "causal skeleton" hypothesis. Bergson's still earlier argument for vitalism is at once a focus of criticism and a helpful means of positioning my attack on determinism; and Popper's more recent account of "downward causation" is germane to my efforts to advance the discussion of issues about levels of emergents and interlevel event determinations. Of course borrowed tools can occasionally prove to be unsatisfactory, and so it becomes necessary to get others. I am inclined to think that discussions of physical determinism have tended to be bogged down by excessive reliance on some inadequate, overworn logical tools; so before I set about defeating the case for determinism, I adopt and adapt some conceptual resources from mathematical analysis to propose a friendly reformulation of the most fundamental argument for the causal skeleton, which I explicate and identify with physical determinism.

Of course not all changes in philosophical currents are improvements, and some of the important auxiliary objects of my critique of instinct—namely, psychological representationalism and computational models in cognitive science—are more or less the received opinions in contemporary philosophy of mind. My attacks against the latter are rather explicit in the recurrent critical discussions of computer simulations, artificial intelligence (AI) theorizing, and psychological functionalism. My antipathy to the former may be somewhat less readily discerned: I mean tacitly to endorse Dewey's nonrepresentational account of experience; to insist—in a roughly Wittgensteinian way—upon the inadequacy of nonprehensile representationalist theories of intelligent functioning; to avoid excessively representational accounts of nonpurposive teleology; to reject the idea that intentional behavior presupposes internal linguistic representations—and to deny instinct. A case could be made for the idea that instinct positing involves ontologically excessive realizations of representational structures. The oddity of my position is that, despite my wish to limit the role of representations, I introduce the notion of some arguably representational abstract structures, teleological schemes, and proceed to suggest that they are ubiquitous accompaniments of animal behaviors. My principal defense, on the supposition that teleological schemes are representations of some sort, is to say that a scheme finally proleptically realized, the abstract structure of an intention qua proximate cause of action, need not be among the organism's original array of schemes; but the meaning of this antirepresentationalist, antinativist remark will not be fully intelligible until much later in the text. Suffice it to say for now that I am mainly against multiplying representational realizations beyond necessity, and that I find much contemporary theorizing in "cognitive science" and philosophy of mind to be insufficiently parsimonious.

My discussion ranges critically over a number of distinct areas of scientific and philosophical inquiry, yet I hesitate to claim interdisciplinary status for my enterprise. My perspective is thoroughly philosophical, and my positive account might not even qualify as "naturalized." The worry is that my approach is too far removed from what is usually done under this last rubric. One might suppose that naturalizing has something to do with the science of biology, and I am engaged in a sort of "biologizing" of philosophical psychology. But that doesn't seem to fit the prevailing notion of naturalizing. Computer science is what many naturalizing theorists seem to associate with their enterprise, and this field seems to me less a science than a theoretical branch of technology. An additional reason for me to avoid this trendy appellation, though, is that I expend much effort in going argumentatively against nature, in one way or another. It would be misleading, then, to describe myself as naturalizing anything—with the possible exception of nature itself.

Biologists will surely recognize substantial portions of my discussion as philosophy of biology, but I doubt whether many of them would concede that my work has very much in common with basic biological research. I would concur; for although there are many issues of overlapping interest between philosophy and biology, there are some central differences between them as disciplines. The trouble with a lot of interdisciplinary work is that it lacks a discipline—unless it manages to carve out a new one. I am largely content just to test some of the limits of defensible philosophical intrusion in the realm of biobehavioral sciences.

There are, of course, some interesting possibilities of disciplinary cross-fertilization—take, for example, von Neumann's account of "self-reproducing automata," which is discussed in section 11.4. This abstract, formal account has the makings of a sensible resolution of the scientific and philosophical worry that evolution seems to involve per impossibile getting something more complex out of something less so. And still closer associations among disciplines are possible, too: I suggest that, in the case of psychology, there are excellent prospects of a true disciplinary (re)union with philosophy—one based on a philosophical type of psychological methodology.

Regardless of its disciplinary affiliations, my primary task has been to try to demythologize the study of behavior, to dispense with that transformed Greek fatalism that now locates behavioral destiny not in the hands of Olympian gods but in our genes. This newer mythological perspective is still more fatalistic than its predecessor, which at least conceded some range of genuine control to human beings and forced the gods to scramble if they ever hoped to oppose and thwart that control. The modern view spurns the tragic drama of the Greeks—by internalizing the ulterior forces that dictate our moves upon the stage, by turning our self-awareness of control into some farcical misunderstanding on the part of "lumbering robots."

My skepticism is an effort to recover our animal dignity. Unlike a proper Pyrrhonian, I do not profess indifference between the skeptical options: I am against instinct, and I am a confirmed believer in the existence of flexibility. My alternative to nativism is proffered not merely as a possible reason for doubt but as an attractive, philosophically compelling conception of a choice region of reality—home to many members of the animal kingdom.

That the new myth, the contemporary version of fatalism, could get started on an allegedly scientific basis is cause for some concern but should not lead us to despair about our meager allotment of reason and understanding. By adopting a critically alert stance in relation to the cognitive and behavioral sciences, philosophy can expose their misunderstandings. Philosophy, our passion to comprehend, will thereby testify to a greater capability of our intelligence than its—than our—fatalistic blunder would suggest. But if we are ever to recapture the prize of our own animal nature, we must straightaway embark against prevailing forces; for Artemis, if I judge her character correctly, does not really wish us to begin by sacrificing, in her name, too much.

Acknowledgments

Words can probably express my gratitude to those who helped with this book, but I am not so sure that I can. Karen Hanson's help has been peerless, matched only by her own previous philosophical and personal succor. Karen is a relentless, remorseless critic of philosophic writing, so she has often been almost too helpful; but I do, sooner or later, have at least a grudging appreciation for this—as perhaps we all should have when significant, intelligent others take our words and our ideas seriously. There is sometimes a fine line between acrimonious and philosophical argument, and Karen has managed to preserve it even when her fellow conversationalist has been less scrupulous. By the way, I love her madly.

During the seven-plus years it has taken to bring this project to fruition, Israel Scheffler has been a source of much friendly encouragement and wise counsel. I am indebted to him too, as so many others are, for the most admirable example he sets of personal and philosophical integrity, of wonderfully sane and judicious thinking, and of other intellectual virtues too numerous to list. I have had the benefit of a supportive, knowledgeable editor, Jane Cullen; and, in consequence, I have also been fortunate enough to get some generous, thoughtful, and helpful referees, whose identities are now known to me: Peter H. Klopfer, whose own work in areas related to the ethological topics of this book is quite renowned, has graciously provided me with a list of references to help in supplementing my discussions of various points. Although I have not, in the space and time available to me, been able to incorporate due considerations of all the views to which he referred me, I have tried briefly to note their relevance to my enterprise. And Donald F. Gustafson, who is, fortuitously, sympathetic to a surprising number of the philosophical intuitions (about mind and action) that prompted this book, has made many constructive suggestions about philosophical issues that bear further consideration. Many, alas, still do, but various adjustments in the text have been made in order to indicate possible approaches to, if not resolutions of, some of the issues he mentioned.

I have profited from discussions that have ensued when I have presented portions of this work in philosophical forums: "How Not to Identify Innate Behaviors," at the 1986 Philosophy of Science Association meeting; "Consciousness Naturalized," at a 1989 colloquium of the Indiana University Philosophy Department; and "Biology, Behavior, and Information Theory," at the 1990 PSA meeting. The first and third have since appeared in *PSA 1986*, vol. 1, and *PSA 1990*, vol. 1, respectively. A version of the second, "Consciousness Naturalized:

Supervenience Without Physical Determinism," appeared in the *American Philosphical Quarterly*, vol. 28, no. 1, January 1991. I thank the editors of these publications for their permissions to reprint material from those papers in chapters 1, 2, and 10 of this book.

Work for this book began in a sabbatical leave granted me at Indiana University, in 1983–1984. In 1985, the university's Office of Research and Graduate Development awarded me an Emergency Grant-in-Aid of Faculty Research, with matching funds from the School of Education Research Initiatives Account, to facilitate the preparation of the manuscript. Sandra Churchill, Personnel Systems Manager, helped guide me through a nightmarish maze of word processing, and, through her offices, a vast army of secretaries, but especially Sandy Strain, typed and endlessly reprocessed revisions of the original deluge of handwritten pages. I will not thank any hardware, software, or my genes.

Part One

The Scientific Quest for Instinct

Biology has no greater triumph to look forward to than a solution of the problem of how a program of instinctive behavior is genetically stored and epigenetically retrieved.

—P. B. and J. S. Medawar

Chapter 1

The Slippery Notion of Behavioral Innateness

Questions have of late been raised about how certain social, altruistic instincts and innate structures underlying linguistic behavior might be reconciled with biological theories of evolution. But, reverting to an ancient stratagem, one might reasonably refuse to take up these questions until provided with some satisfactory answers to prior questions about just what is meant by innate components of behavior and (a less Socratic question) what good direct or indirect grounds one might ever have for positing innateness with respect to behavior. Part One of this book explores these prior questions within a broadly scientific context. Special attention is paid to the points of contention between pro-instinct ethologists and anti-instinct comparative psychologists, two groups that agree in favoring experimental approaches to the study of behavior; but other, more theoretical approaches to instinct are considered, too.

1.1 Snarled Criteria of Innate Behavior

The first place we might think to look for our own instincts would be among the behaviors exhibited by newborn babes. According to Jean Piaget, the infant's original behaviors include "sucking and grasping reflexes, crying and vocalization, movements and positions of arms, the head or the trunk, etc." [1] He contends, not implausibly, that these behaviors are "inherent in the hereditary equipment of the newborn child." [2] But what makes this so obvious? On what grounds are we ever entitled to conclude that specific behaviors are instinctual?

Philosophers have raised related questions about the innateness of ideas (rather than behavior), and, if the history of this debate is any indication, the prospects for finding such grounds in the case of behavior are slim. Typical philosophical objections to innate ideas are leveled against suggested criteria of innateness. Thus, John Locke argues against one proposed criterion (viz., "universal consent") as follows: "If . . . there were certain truths wherein all mankind agreed, it would not prove them innate, if there can be any other way shown how men may come to that universal agreement." [3] And an analogous criterion of innateness in the case of behavior has been suggested by, among others, Charles Darwin:

"Guanocoes have the habit of returning . . . to the same spot to drop their excrement . . .; as this habit is common to all the species of the genus, it must be instinctive."[4] So Locke's argument would seem apposite: The habit might, for example, be learned by all the Guanacoes and hence need not be instinctive.

The possibility of learning seems such a natural alternative to the hypothesis of innateness that one might be tempted to characterize the innate as the unlearned; but, as D. O. Hebb reports, "the chemical environment of the mammalian embryo, and nutritive influences on the invertebrate larva, are factors in behavior which do not fall either under the heading of learning or under that of genetic determinants."[5] If innateness implies hereditary determination, then such chemonutritive factors might yield behaviors neither learned nor innate.

One might think simply to characterize innate behavior as that which is determined by heredity. But if, as Hebb supposes, "no behavior can be independent of an animal's heredity,"[6] then this characterization would have all behavior qualify as innate. Hebb insists his supposition "is so obvious, logically, that it need not be spelled out." Is it? Only if we *pre*suppose a rather mechanistic physiological conception of the behaving organism. Were we, instead, to credit the idea of physiologically undetermined behavior, of radically free action, Hebb's supposition would lose its force. Of course this rejoinder may seem excessively academic since Hebb's presupposition is so widely shared; and, besides, even if some behaviors are noninnate in consequence of being thus undetermined, that still seems to leave too many other behaviors qualifying as innate. Hebb insists that "learned behavior . . . can never be thought of as something apart from the heredity that made possible a particular kind of sensory structure and nervous system."[7] But this would make even learned behaviors, determined by heredity, innate.

There is, it should be noted, some ambiguity in this use of the phrase "determined by heredity": It might mean that the character of what is learned is dependent upon heredity, or it might mean that some sort of mechanism of learning is part of one's hereditary endowment. These interpretations need not be mutually exclusive, since the character of what we learn might, arguably, depend upon the nature of our inherited learning mechanism (if such there be). But if learning is said to be determined by heredity only in the latter sense of that phrase, then maybe learning can nevertheless be viewed as *not* being determined by heredity in the former sense of that phrase. Assume for the sake of argument that something like B. F. Skinner's theory of operant conditioning is at least partly correct, that some unpredictable, spontaneously emitted behaviors are learned when rewarded or "reinforced" in various ways. Those behaviors may be reinforced in ways fully dependent upon the organism's inherited learning mechanism, and yet their character may not be causally determined at all, much less by heredity. As Skinner puts it, "the operant response is 'emitted.' . . . The principal feature is that there seems to be no necessary prior causal event."[8] And a convenient physical analogy to this feature may be found in the emission of subatomic particles by radioactive substances, a process that "is at present not known to occur only under specific determining conditions."[9]

Care must be taken, however, not to overstate the suggestion that learned behavior might be genetically undetermined. Just as, for all the apparently undetermined character of particle emission from a piece of radium, we should not expect to discover those emissions to be very tiny baseballs, so too, for all the apparently undetermined character of operant behavior, we should not expect that behavior ever to consist, for example, of human beings' flying by means of the rapid flapping of their naked arms. The suggestion that the character of learned behavior might, conceivably, be undetermined by heredity is consonant with the admission that the range of that behavior is severely limited by the genetically (and environmentally) determined character of the physical organism. The things we learn to do must be things we are (physically and otherwise) capable of doing; but this truism does not imply that our spontaneous doing of these things is genetically determined.

Hebb in fact appeals to the alleged hereditary determination of learned behavior as one of two supports for his overall contention that it is a mistake to attempt to sort all behavior into two disjoint categories, learned and innate. His contrary suggestion is: "We are on solid ground if we think consistently of all behavior as 'caused by' or fully dependent on both environment and heredity, and cast our research in the form of asking how they interact."[10]

Hebb's other support for his overall contention is this: "I . . . refer you to the finding that mammalian perceptions in general appear to depend not on formal training, it is true, but on a prolonged period of patterned sensory stimulation. . . . All that a mammal does is fundamentally dependent on his perception, past or present, and this means that there is no behavior, beyond the level of reflex, that is not essentially dependent on learning."[11]

As it stands, the argument seems incomplete—requiring the additional premise that "a prolonged period of patterned sensory stimulation" amounts to learning. Since Hebb subsequently defines learning as "a stable unidirectional changing of neural function resulting from sensory stimulation (including the stimulation that results from response),"[12] the requirement is easily met. Hebb's full argument, then, is this: All (nonreflex mammalian) behavior is dependent on perception; all (mammalian) perception is dependent on learning; so, all (nonreflex mammalian) behavior is dependent on learning.

Hebb seems convinced that his conclusion undermines the distinction between the innate, qua unlearned, and the learned: "Fear of strangers . . . or a temper tantrum is not learned and yet is fully dependent on other learning. Do we then postulate three categories of behavior, (1) unlearned, (2) unlearned but depending on learning, (3) learned? Perhaps instead we had better re-examine the conception of unlearned versus learned behavior."[13]

But perhaps one such rhetorical question deserves another: Why *not* postulate these three categories of behavior? Hebb's answer would likely be that since innateness is characterized in terms of nondependence upon learning, the category of unlearned but dependent on learning would—within the terms of this controversy—be incoherent, that this category would amount to one of behavior that is innate and yet not innate. But it may be that what is problematic is Hebb's own tacit understanding of innate behavior as behavior not dependent upon learning. A more proper understanding of innate behavior would be that it is behavior that is, among other things, not itself learned. Hebb's apparent failure fully to appreciate the importance of the difference between these divergent accounts of innate behavior might well be attributed to his above-mentioned account of learning (as "stable unidirectional change of neural function"); for this account has a tendency to mask that difference by ignoring the content of what is learned in favor of emphasizing the physiological counterparts of learning.

This masking phenomenon may be made clearer by way of the following cases: Hebb himself says that the shyness of chimpanzee babies "is not learned: but it is definitely a product of learning in part, for it does not occur until the chimpanzee has learned to recognize his usual caretakers." [14] But if we ask whether, given his definition of learning, the shyness has been learned, then the answer is yes, since the shy behavior would be presumed to be a consequence of a "stable unidirectional change of neural function." And the same answer could be given, on the same grounds, to the question of whether a child's shoe-tying behavior is learned. So Hebb's own initial willingness to countenance a distinction between learning-dependent behavior and learned behavior does not easily survive his modest reductionism, his physiological definition of learning: The closest reductionistic parallel to that distinction might be one between consequences of the neural changes and those neural changes themselves; but since we cannot directly identify behaviors with neural changes, this parallel is not close enough to ground the distinction between learning-dependent and learned behavior.

Given the apparent inability to ground this distinction, one might begin to wonder whether the further distinction between learned and innate behavior can intelligibly be drawn at a purely physiological level of discourse. If it cannot, then there is no real hope that future developments in the field of physiology will resolve every question about whether a given bit of behavior is learned or innate. And no solace may be found in the notion that, once it is determined that there is no purely physiological counterpart to the distinction between learned and innate behavior, these questions about the proper classification of particular behaviors can be set aside as pseudoquestions about an illusory distinction. For even if that distinction cannot be drawn at one level of discourse, it does not follow that it cannot be drawn at any other level.

Such cautionary remarks as these are premature however, since it has yet to be determined that the innate/learned distinction cannot be drawn in purely physiological terms. Even Hebb's (tentative) physiological definition of learning doesn't rule out the possibility of drawing any such distinction—it merely raises a question about whether some intuitively instinctive behaviors (those that are learning-dependent but not learned as such) should not be otherwise classified. Of course if, as Hebb suggests, all (nonreflex) behavior is learning-dependent, then the whole distinction is in danger of collapse: Given a breakdown of the learned/learning-dependent distinction, no (nonreflex) behavior could be classified as being innate as opposed to learning-dependent (and/or learned).

1.2 A Forthright Experiment, Some Possibly Crooked Facts

Konrad Lorenz, a major defender of the distinction between "ontogenetically acquired" and "phylogenetically adapted" behavior,[15] suggests that adequate grounds for classifying some behaviors as innate are to be found in the results of what he calls "the deprivation experiment": "the experiment of withholding from the young organism information concerning certain well-defined givens of its natural environment."[16*] Thus, a stickleback fish is deprived of the information that its rival has a red belly. The stickleback is then confronted, for the first time, with a red-bellied rival (or a red-bellied dummy). If that stickleback responds with species-typical rival-fighting behavior, then (according to Lorenz) the experiment has established that the stickleback possesses certain innate information about its natural environment. On the other hand, should the stickleback fail to respond in this way, Lorenz tells us that

> we should not be justified in asserting that this response is normally dependent on learning. There would still be the alternative explanation that, while trying to withhold from the animal information only, we have either inadvertently withheld "building stones" indispensable to the full realization of the blueprint contained in the genome or else we are withholding in the experimental setups a stimulus situation necessary to release the behavior we are undertaking to investigate.[17]

A corresponding difficulty might seem to stand in the way of inferring innateness from the first-mentioned result of the deprivation experiment, since we might have failed to withhold information required for the stickleback to learn to recognize its rival. Lorenz may be aware of this difficulty, but he seems to regard it as surmountable: He warns us against the possibility of "extremely quick 'flash-like' conditioning."[18] Indeed, this possibility suggests to him a difficulty with the stickleback experimentation: "If on this first occasion of presentation, the dummy was red below, it is quite possible that the innate information is confined to 'red' alone and that the configurational property 'below' is learned in a flash."[19] But notice that this example only partly confounds the deprivation experiment, by suggesting that not all the information is legitimately inferred (from that experiment) to be innate. And note, too, that the stickleback is presumed to have been deprived, until the eventual presentation of the dummy, of the information that allows for "learning" (i.e., for rapid conditioning).

Yet withholding pertinent information may be more difficult than Lorenz supposes. Consider the fanciful possibility that the stickleback's mother had told him long ago, "Beware of the red-bellied fish—he's your rival!" Without taking this example seriously, we may still draw from it a serious lesson: Learning about something does not necessarily depend upon one's being presented with that something.

Furthermore, and no doubt much more relevant to the case of sticklebacks (as opposed, say, to human beings), Lorenz's experiment must not only deprive the organism of any possibility of learning the information to be tested for innateness, it must also rule out any other way of acquiring that information. Is there some third (serious) possibility besides innateness and learning? Much depends, obviously, upon what we mean by learning. Given Lorenz's own willingness "to define learning in a much too comprehensive way, [as] including all adaptive [ontogenetic] modifications of behavior," [20] perhaps many initially plausible candidates (e.g., chemical reactions with environmental agents) will end up being classified as cases of learning, since anything that causes the stickleback to react aggressively toward its rival could readily be interpreted as an adaptive modification. But while this may rule out any third possibility, Lorenz could not take any comfort from that fact: The more that counts as learning, the more that the organism must be deprived of in such an experiment.

Without sufficient assurance that all the right factors are eliminated, the deprivation experiment can no more prove that a behavior is innate than it can prove that other behavior is learned. Still, it is a mistake to think that the deprivation experiment presupposes the possibility of determining which, if any, behaviors are learned. That experiment demands only that the experimenter be able to identify essential prerequisites of possible learning. Thus, in the familiar case of the stickleback, the experimenter is called upon to make the following sort of conjecture: If the stickleback is to learn to identify red-bellied members of the same species as rivals, then either he must be directly confronted by such a rival or his mother must inform him or he must see red-bellied conspecifics act as rivals toward other male sticklebacks or. . . . The list may go on, but is there any reason to suppose that it might go on indefinitely? Such a list, reasonably if not logically complete, is all one needs for the successful exploitation of the deprivation experiment. And should one ever fail to deprive the organism of a likely occasion for learning, the appropriate way to correct the mistake would be a further application (not the abandonment) of the deprivation technique. So we have yet to diagnose any fatal theoretical problem inherent in the deprivation experiment.

There is, though, a problematic bias, about the nature of behavior, implicit in Lorenz's discussion of this technique. For although he acknowledges the possibility that a poorly devised deprivation experiment can lead to a pathological disintegration of the "subject's whole system of actions, or 'ethogram,' " [21] Lorenz seems confident that, as long as care is taken to deprive the subject of only the specific information that might be requisite for learning, such "diffuse damage" will be prevented and only "circumscript defects" will be produced.[22] This confidence betrays an atomistic conception of the behavior targeted for study, a substantive presumption about the existence of certain units of behavior. For if the targeted bits of behavior were not virtually independent of one another, one could not expect to interfere with one while leaving unaffected all the others in the organism's ethogram. Target behaviors that were not appropriately independent could not be demonstrated to be innate by means of a deprivation experiment; so if behaviors are typically very interdependent, the deprivation experiment will typically not work.

Other possible facts about animal behavior would also deprive the deprivation experiment of much merit. Suppose, for example, that many animals that learn certain behaviors essential to their well-being also have, as backup systems, innate capacities to perform the same activities. If so, the deprivation experiment would yield the incorrect conclusion that those behaviors are, in the normal environment, innate. Or suppose, instead, that learning serves as a backup for instincts: Then the deprivation experiment might yield the typically correct conclusion that a certain behavior pattern is innate; but, in any given case of that behavior as exhibited in nonexperimental circumstances, one wouldn't know (at least not on the basis of the deprivation experiment) whether that behavior is learned or innate.

1.3 The Intimacy of Learning and Genetic Decoding

Although some of the preceding possible facts were quite contrived, propounded for no other reason than to create difficulties for the deprivation technique, there are other, no less bothersome possibilities that are independently plausible. Thus one might suppose—as Piaget does—that some reflex behaviors are modifiable by learning. If so, the deprivation experiment's usefulness in identifying them as innate would drop off once learning began. And learning might begin very early: According to Lorenz, Daniel S. Lehrman "gives serious consideration to the assumption that a chick could learn, within the egg, considerable portions of the pecking behavior by having its head moved rhythmically up and down through the beating of its own heart." [23] If this assumption is correct, then the utility of the deprivation experiment approaches a vanishing point.

Lorenz concedes that some learning may take place before hatching or birth, but he rejects any suggestion that the embryo could thereby "acquire information on environmental givens which are only encountered in later life." [24] Lehrman, in a subsequent discussion, points out that he had not even used the term *learning*. His intent was "to convey the idea of interpenetration between the processes of growth and those of the influence of environment, and to express a feeling of tentativeness and ambiguity about the distinction between the effects of experience on a developing organism and the effects of experience in a mature nervous system."[25] Rather than speak of learning, Lehrman employed T. C. Schneirla's notion of experience: "the contribution to development of the effects of stimulation from all available sources (external and internal), including their functional trace effects surviving from earlier development."[26]

One can appreciate the plausibility of Lehrman's position: Experience (thus understood) probably does play an important role in the embryological development of many animal species, and it would be difficult to separate out this role, all the varied effects of experience, from the role of processes of growth or maturation. Yet one can also appreciate Lorenz's position, for if learning about the external environment cannot antedate the organism's first exposure to that environment, then it might still be possible to use the deprivation experiment to identify innate aspects of behavior at the time of birth or hatching, before learning could intervene and spoil the experiment.

Of course strictly speaking, as already observed, the general rule that learning something demands direct environmental exposure to that something is incorrect, but conditions that might provide exceptions to this rule would not seem to obtain in the present circumstances: No one is inside the womb or egg to inform the developing embryo about environmental contingencies that the embryo has yet to encounter. And a still more general rule, with which Lorenz seems to agree, would appear to be without exceptions: Learning about something demands an accessible source of information about that something.

1.4 Information as the Innate Component of Behavior

Lorenz does not view behaviors as innate; he does not even regard differences among behaviors (of different species) as innate. Rather he construes information (about the environment to which the behavior is adapted) as the innate component of behavior. The deprivation experiment is intended to withhold environmental sources of that information from the organism: Should the organism nevertheless exhibit behavior evidencing possession of such information, then that information must be innate. Lorenz interprets this conclusion to mean that the information is transmitted to the organism genetically, by way of a sort of blueprint in the genome.

Lehrman quarrels with the figure of a blueprint:

> A blueprint is isomorphic with the structure that it represents. . . . It is *not* true that each structure and character in the phenotype is "represented" in

a single gene or well-defined groups of genes; it is *not* the case that each gene refers solely or even primarily, to a single structure or character; and it is *not* the case that the topographical or topological relationships among the genes are isomorphic with the structural or topographical relationships among phenotypic structures to which the genes refer.[27*]

But perhaps the quarrel is not a major one, since Lorenz himself does contend that this blueprint needs to be "decoded," and the environment, whether internal or external to the womb or egg, could be conceded by him to play a crucial role in the decoding process.

Of more moment than Lorenz's infelicitous figure of a blueprint is his strong overall allegiance to the notion of information. Lorenz never seems to question the idea that certain behaviors demand the organism's possession of information about the environment. But what possible grounds are there for this idea?[28*] The concept of behavior does not itself entail that of environmental information, though some stipulative definitions of the term *behavior* may come close to doing so. Thus, Piaget defines behavior as "all action directed by organisms toward the outside world in order to change conditions therein or to change their own situation in relation to these surroundings."[29] And it might be argued, perhaps not conclusively, that directing action upon something demands some degree of information about that something. However, the term *behavior* is commonly used to signify observable muscular or glandular responses of an organism, and such responses would not generally imply information about the environment: Consider, for example, the knee-jerk response. Does it entail anything even approximating knowledge of the environment? If not, then Lorenz would seem oddly uncommitted to that reflex's having any innate components. And if the term *behavior* is used somewhat more broadly as "any objectively observable activity of the organism," then the possible connection to specific information about the environment seems all the more tenuous.

Although not all behaviors involve environmental information, perhaps some of them do, and, provided that it is possible to identify the latter, they might help vindicate Lorenz's viewpoint, lending themselves to further study via his deprivation technique. Some behavior does exhibit a kind of knowledge, and instinctive behavior, in particular, is said by Piaget, to exhibit a certain savoir faire. He contends that we may investigate what an animal instinctively " 'knows how to do' in response to stimuli perceived in the external environment" and that " 'to know how to' [savoir faire] is a kind of knowledge [*connaissance*] or ability or [savoir] like any other, and in the human child it precedes conceptual knowledge by a wide margin." (*sic*)[30] In the more Anglo-Saxon idiom of Gilbert Ryle, this point might be put by saying that instinctive behavior involves "knowing how" rather than "knowing that."[31] But the question still remains whether such "knowing how" entails knowledge of, or information about, the environment.

Consider the following example of some fairly simple, presumably innate behavior: the pecking behavior of newly hatched chicks. The chicks that exhibit such behavior, might, it seems, correctly be described as knowing how to peck. And this knowledge is of obvious importance to their need for food. But what sort of information about their environment does this pecking behavior involve? It might be argued that since these chicks are pecking for food, they must have some information about their environment—namely, that it contains food for pecking. But are we entitled to infer from their behavior that they are pecking for food? If this means that they are aware that nourishment is a desired consequence of their pecking, then our inference is dubious. If it merely means that the pecking does in fact fulfill such a function, then it is a harmless supposition, but it scarcely warrants the further conclusion that information about the environment is involved. Moreover, even this harmless supposition needs to be qualified carefully, since that same behavior might, in the recent past, also have fulfilled the function of hatching from the egg. Generally speaking, we are not justified in concluding from the fact that an animal knows how to do something that he, she, or it possesses information about the environment: Whistling and thumb twiddling come to mind as but two of numerous activities that involve "knowing how" without specifically requiring information about the environment.

Lorenz's own position does not seem to be that the involvement of specific environmental information is obvious in any given case of knowing how to do something, but rather that various deprivation experiments will disclose what innate information, if any, is involved in the species behavior. To the extent that he does seem to presume a connection between a behavior and the possession of environmental information, that presumption is based on the apparent adaptedness of the behavior to the environment; it is not based on the savoir faire entailed by the behavior.

1.5 Adaptive Behavior, Sign Stimuli, and Information

Suppose that some fictitious chicks, newly hatched, peck at only one very specific kind of object: a piece of grain or a reasonable facsimile thereof. Suppose, too, that deprived of such objects, the chicks do not peck. May we then reasonably conclude that they know whereat they peck, that they are equipped with innate information about their environment? Deprived of a hammer or other blow to my knee, I do not exhibit the knee-jerk response. Is it safe to conclude that I know something about hammer blows, even before the first time my knee jerks in response to such a blow? It may be said that this case is different from the case of pecking because perception provides the stimulus to the pecking response, whereas a more primitive sort of sensory stimulation elicits the knee-jerk response. But it seems strange to have our decision about whether a response is innate depend upon its being elicited by perception, especially if perception depends upon learning.

A more promising account of the differences between these cases requires further details from Lorenz's theory. Following W. Craig's terminological lead,[32*] Lorenz distinguishes between "consummatory behaviors," which are often, like pecking, of direct vital importance to the organism's biological systems (e.g., the ingestive system) and "appetitive behaviors," which are, like the chick's walking and scanning the ground before itself, purposive behaviors, directed toward the attainment of the consummatory behaviors. A consummatory act involves an "Erbkoordination" or fixed motor pattern, which is, by Lorenz's reckoning, "the Archimedean point on which all ethological research is based."[33]

Thus, the chick may be said to have information about appropriate targets for pecking, because that pecking is the goal of appetitive behavior, whereas my knee-jerk reaction is not such a goal. This claim is a little obscure, but the point might be clarified by saying that, relative to the organism's purpose of pecking for food, pecking at grainlike objects is evidence of the chick's possession of information about the environment. The consummatory behavior of pecking, adapted as it is to the environment, requires information about biologically appropriate circumstances for its occurrence. That information, not the fixed behavior pattern itself, is what Lorenz calls innate.

But this account may still not be entirely satisfactory. Suppose I go to a doctor, on a fairly regular basis, in order to keep various of my biological systems operating properly. I might be said to engage, then, in the appetitive behavior of seeking out and visiting a doctor to have my reflexes tested, to exhibit a knee-jerk response. This response might be construed as consummatory behavior. As such, it should, but does not, involve information about the environment: I might not even know, on my first visit, or remember, on later visits, the particular fact that my reflexes will be tested in this way.

But perhaps some appetitive-consummatory behaviors always do involve information about the environment. Consider appetitive behavior that terminates in the perception of what Niko Tinbergen calls a *sign stimulus,* which elicits or releases a fixed behavior pattern. For instance, the red spot on its parent's bill is a sign stimulus for the herring gull chick: The stimulus releases the chick's behavior of pecking at the spot, causing the parent to regurgitate food and feed it to the chick.[34] Given the notion of a sign stimulus, it seems easier to speak of information-laden behavior. Thus, the herring gull chick who responds to the sign stimulus just mentioned might be said to possess the information that the place that must be pecked is the red spot. And, the stickleback who responds aggressively to the sign stimulus of a red belly might be said to possess "the information that the rival who must be fought is red on the underside."[35] On the other hand, it is not so plausible either to regard a blow to the knee as a sign stimulus or to say that this reflex behavior involves information about the environment. The blow is not a sign of anything, so the reaction to it need not involve any information about the environment.

Intuitively, then, some difference obtains between fixed behavior patterns released by sign stimuli and simple reflexes elicited by direct stimulation, but the precise basis for that distinction is still a mystery. The greater complexity of a sign stimulus might help to distinguish it from a nonsign stimulus, but such a characteristic seems only accidentally relevant to something's being a sign, the more essential characteristic of which is to indicate something. The adaptive character of the consummatory behavior released by a sign stimulus might seem a better clue to the difference between sign stimuli and ordinary stimuli (of simple reflexes). Thus, the sign might be taken to indicate that its environment is (more likely to be) "right" for the released behavior, that in such a setting the behavior would tend to promote the animal's survival. But this clue is far from conclusive, since even the knee-jerk response might once have been a valuable defense, say, against predators who began their assault by striking our ancestors' knees.

The connections among adaptive behaviors, sign stimuli, and information about the environment may not be as straightforward as they seem: Suppose that a particular species has, during its long phylogenetic development, adapted to a particular environment but that some cataclysmic events have so transformed the environment that this behavior is no longer of any real survival value. Is the stimulus that elicits this no-longer-adapted behavior a sign stimulus? If we decide to say that it is, should we also insist that this species is thus shown to possess information about its environment? The heroic course might be to suggest that the species now possesses incorrect information about its environment, but such heroism may lead to further difficulties down the road: Suppose the environment changes once again—not back to its original state, but to some novel state in which the stimulus serves to elicit the same behavior, fortuitously fitting once again. For example, a chick might once have pecked at what was then its only available food source, tiny red beetles; and the same species might now peck at what is presently its only available food source, little red seeds. Surely it would be overly heroic to contend that the chick presently has incorrect information about its environment; yet the temptation to say some such thing remains. After all, is it not the chick's innate information about tiny red beetles that leads it now to peck at the right sort of thing for the wrong reason?

A more reasonable course would be to steer clear of any such appeal to the notion of information. The relevance of sign stimuli to vital occasions for certain behavioral responses may be far too coincidental, the evolution of behaviors much too chancy, to warrant the idea that phylogenetically adapted behavior involves real information about the environment. But since a temptation to speak of information remains, the latter's putative role bears further examination.

1.6 Evolution and Information Theory

Let us consider what further light might be shed on Lorenz's use of the term *information* by that term's special technical acceptation in communication theory.

The essential features of a communication system are these: "The original sender selects a message out of the *message source*. . . . This is then *coded* by the *transmitter* . . . into the *signal* . . . which pass[es] over the *communication channel* . . . to reach the *receiver* where it is *decoded* . . . eventually into the *message* which then passes to the *destination*. And we must note that *noise* (a general term for undesired additions to the signal) may have entered . . . so that the *received signal* differs from the originally *transmitted signal*." [36*]

Now the "great book of nature" might, for argument's sake, be viewed as a message source, but the mechanisms of neo-Darwinian evolution can hardly be viewed as a communication system that transmits information from that source to successive generations of organisms. The discoveries of molecular genetics have led to a well-developed biological theory in which the communication theoretic notion of information does play a prominent role: "A gene is now known to be a segment of one of the extremely long DNA molecules in the cell that store the organism's genetic information in their structure. The sequence of four kinds of nucleotide base . . . along each strand of the DNA double helix represents a linear code. The information in that code directs the synthesis of specific proteins; the development of an organism depends on the particular proteins it manufactures." [37]

But this communication system does not receive information from the environment. Indeed, although some environmental input (e.g., ionizing radiation) may give rise to mutation, the primary engine of evolutionary changes,[38*] that input is most straightforwardly interpreted as randomly generated noise, not as information selected, encoded, and transmitted from that environment. And since no information *from* the environment is ever transmitted via molecular mechanisms of evolution, to successive generations of organisms, it is reasonable to conclude that those evolutionary mechanisms do not give rise to any real information *about* the environment.

This argument rests on an assumption that information *about* something must have that something as its source, must be information directly or indirectly *from* that something. And it might be objected that this assumption runs afoul of hypothetico-deductive strategies of learning about the world. After all, scientific theories are hypotheses that have the scientist as their source but that are, nonetheless, about the world. Moreover, it might be suggested that evolution itself proceeds by way of some such strategy, one in which adaptation rather than scientific explanation is the goal: "The modern view of adaptation is that the external world sets certain 'problems' that organisms need to 'solve,' and that evolution by means of natural selection is the mechanism for creating these solutions." [39*] So, it might be urged, a species may derive information about the environment from random mutations, from a source endogenous to individual organisms belonging to that species. And, once these mutations afford tentative solutions (or hypotheses), then further information (in the form of negative feedback) may be forthcoming directly from the environment.

However, the only reason hypothetico-deductive strategies in science afford information, via negative feedback, about the environment is that the hypotheses generated by scientific communities are intended for that purpose, are directed from the start toward that goal. This rejoinder may not seem very persuasive; for, to put one likely objection figuratively, so long as one casts one's net upon the waters, it doesn't much matter whether one is trying to fish. But notice that even before these hypotheses are put to any test, before there is any occasion for negative feedback from the environment, scientific hypotheses are already properly described as being *about* the environment. Nothing comparable may properly be said about the "noise" that originates in random mutation and that, having been transmitted to successive generations, occasionally proves useful to the species in its struggle for existence: Mutations, even ones that promote adaptation to the environment, cannot be said to involve information about the environment.

These remarks are presented in a neo-Darwinian spirit and may not accord equally well with other theories of evolution. If, for example, Lamarck were correct in supposing that acquired characteristics might be biologically inherited by subsequent generations, then it might make sense to speak of an organism's selecting information from and about the environment and transmitting that information to its offspring. But the ethological evidence for Lamarckian theory is both scant and disputable, and no direct support for its tenets seems forthcoming from other biological sciences. Furthermore, the sort of Lamarckian hypothesis that seems favorable to information-theoretic accounts of instinct raises an obvious question: what strange channels of communication could possibly convey specific environmental information from its locus in the brain, where it has been said to be encoded "into a spatial and temporal pattern of nerve impulses,"[40*] to the DNA contained in the seemingly isolated germ cells of the organism?

Lorenz himself is so intent upon avoiding Lamarckianism that he rejects the hypothetico-deductive model of adaptive processes: "It is only individual learning that is able to gain information from its errors. The hit-and-miss method of mutation and selection gains only by its successes and not by its failures and continues blindly to produce those mutants that have proved unsuccessful millions of years ago."[41] Given this rejection of the analogy between trial-and-error learning and mutation-and-selection adaptation, there is little reason to view evolutionary mechanisms as yielding information about the environment.

But is it really so obvious that mutation is utterly blind to past failures, especially to those that are so maladaptive as to decrease an organism's chances of survival? In a purely speculative way, let me suggest the bare possibility that natural selection might decrease the chances of reoccurrence of maladaptive mutations: The selection pressure against the phenotypic consequences of specific mutations might also be selection pressure against those DNA molecule-types that have a tendency to mutate in the specific ways that have, within the population, proved disadvantageous in the past. In communication-theoretic terms, this hypothesis would involve information *from* the environment and within the genetic communication systems. However, it would still provide no justification for the claim that that information was information *about* the environment.

1.7 Behavioral Images of the Environment

One reason for Lorenz's reliance on the notion of information is his understanding of the idea of adaptation: " 'Adaptation' is the process which molds the organism so that it fits its environment in a way achieving survival. Adaptedness is always the irrefutable proof that this process has taken place. Any molding of the organism to its environment is a process so closely akin to that of forming, within organic structure, an image of the environment that it is completely correct to speak of information concerning environment being acquired by the organism." [42*]

There may be an air of obviousness about the suggestion that a representation (or "image") of the environment amounts to information about (or "concerning") that environment. But what justification does Lorenz have for his prior suggestion that adaptation to the environment involves (or is "closely akin to") representing the environment, that an organism adapted to an environment possesses its own structural representation of that environment? Such terms as *organic structure* and *image* are billowy enough to cover the situation being described (an organism's adaptation to the environment), but those terms cannot be said to help clarify that situation. Consider an analogy: I learn how to ride a bicycle. Does it follow that I afterward possess an organic structural representation of a bicycle? Suppose further that I have been taught by what my instructor likes to call "the touch method," that during the learning process I was fitted with blinkers designed to prevent me from seeing the bike instead of the road. Is it still reasonable to suggest that I must, upon learning to ride, possess an internal representation of the bicycle itself? Suppose there is no bicycle, that I have learned "bicycling movements" in a pool, from a friend hoping to teach me to tread water. Do I still have an image of a bicycle, or is it now an image of the water? Without denying that something must be there, in me, to account for my recently acquired know-how, it is possible to reject the slightly absurd further suggestion that that something must include a specific image of the environment in which my know-how is manifested, a structural representation of the bicycle or of the water. Is it not at least equally plausible to suggest that my know-how is represented organically exclusively in the form of certain so-called motor memories, making no reference whatever to a specific environment?

The skeptical businessman may have a point when he opines that a highly educated MBA might still not know about the real world of business, even if the graduate has, in Accounting 101, learned how to balance a company's accounts. Of course the businessman does overstate the case, since accounting procedures have some reference to, and embodiment in, the business environment; but although learning how to do some things may inevitably involve learning about the environment in which they are done, learning how to do things does not in general entail learning about the environment: Suppose I "learn" the conditioned response of thrusting my fist violently forward whenever some stimulus elicits my unconditioned knee-jerk response. I might then have learned how to do something that in certain environments (not, of course, a doctor's office) has survival value. I do not thereby possess an internal structural representation of those environments; I have not learned about them.

Some of the very features of behavior that are, according to Lorenz, characteristic of "innate information" should be cause for concern that his talk about information is problematic. Thus, Lorenz comments, "It is a fundamental error to think that selectivity of a response indicates the presence of innate information. The exact opposite is true. It is quite a reliable rule of thumb . . . that if the animal is taken in by simple dummies one can assume that the underlying information is innate while, conversely, high selectivity is practically always an indication that learning has played a part." [43] But this lack of selectivity amounts, presumably, to a lack of specific information about the environment to which the animal's behavior is adapted; and a lack of such information is surely no indication that the animal does possess other innate information about its environment.

Another curious characteristic of innate behavior—curious only if one regards information as being what is innate—is the possibility of releasing it by means of what Tinbergen calls "supernormal" sign stimuli, that is, "stimulus situations that are even more effective than the natural situation." [44] Among examples he cites of this phenomenon is "the oystercatcher's preference for abnormally large eggs. If presented with an egg of normal oystercatcher size, one of herring gull's size, and one double the (linear) size of a herring gull's egg, the majority of choices fall upon the largest egg." [45] Now, if the oystercatcher exhibits an innate preference for eggs larger than its own, that bird would seem to have rather queer innate information about its environment. One could almost say that it has information not about its own but about a supernormal environment. Better still, one could simply deny that putatively innate behavior involves information about any environment.

Perhaps part of the temptation to speak of such information is a misleading conception of what it means for behavior to be adapted to its environment. Describing phylogenetically adapted behavior as fitted to its environment can easily mislead us into supposing that random variation and natural selection have combined to adjust the behavior to that specific kind of environment. For all we know, more often than we assume, the adaptation of animals' behavior does not follow such a scenario. Leaving aside the creationist and Lamarckian schemes, there are many credible alternative plots available that are fully compatible with the larger structure of neo-Darwinian theory. These plots emphasize the role of changes, both alterations and alternations, of the environment.

Particular alterations of the species' normal environment are too often, as a matter of intellectual habit, viewed as "problems" whose "solutions" may be provided by the species' adaptive mechanisms. But one might as well suggest that in other cases the changing environments provide "solutions" to "problems" posed by random mutations within the species. And if this metaphoric talk of "problems" and "solutions" leads some theorists to speak of the organism's innate information about the environment, then we might speak, with equal justice, equally absurdly, of the environment's innate information about the organism that thrives in it.

Alternation of the environment occurs when an organism's evolved appetitive behaviors transport it to a different locale, but an alternation may also occur when, say, a stiff wind blows the organism to a novel setting. In the new environment, "sign stimuli" adequate to release consummatory behaviors that promote survival are sometimes available. These behaviors may thus come to fit an environment not because they have been phylogenetically adjusted but because the environment has fortuitously shifted. Even if the alternation results from innate appetitive behavior, the consummatory behavior is adapted without the organism's having undergone evolutionary adaptation, "molding," to achieve the new fit—contrary to Lorenz's contention that the former is "irrefutable proof that [the latter] process has taken place." Once again, talk of innate information about— or evolved behavioral images of—the environment is not apt.

1.8 The Uninformed Elasticity of Supposedly Innate Behavior

If innate behavior is phylogenetically adapted, then perhaps noninnate behavior is adaptive, instead. The adapted/adaptive distinction might be thought to correspond to Tinbergen's consummatory/appetitive distinction: "The consummatory act is relatively simple; at its most complex, it is a chain of reactions, each of which may be a simultaneous combination of a taxis and a fixed pattern. But appetitive behavior is a true purposive activity, offering all the problems of plasticity, adaptiveness, and of complex integration that baffle the scientist in his study of behavior as a whole." [46] However, a taxis (or "sequence of reactions to external stimuli that continuously correct the direction of the movement in relation to the spatial properties of the environment" [47]) could prove difficult to distinguish from "a true purposive activity." And any consummatory behavior involving a taxis could adaptively transport an organism to a novel environment. Moreover, as the phenomenon of supernormal sign stimuli makes evident, there can be a range of stimuli capable of releasing a consummatory behavior. Even the fixed behavior pattern, Lorenz's paradigmatically adapted, innate, form of behavior, can exhibit survival value in a wide range of environments—as in the fictive example of the chicks who once pecked at bugs, then at seeds. So, many seemingly adapted behaviors, even ones that fit their environments "like a glove," may be somewhat adaptive, too.

The bare fact that a given behavior is well molded to an environment is no sure sign that the behaving organism possesses an innate image of—or information about—that environment. Consider the analogy of clothing that is well molded to its wearer. Given the advent of so-called one-size-fits-all garments, which literally mold themselves to the wearer, one could say that some clothing is the very image of anatomical features of its wearer. The structure of the clothing—say, a dancer's leotard—might be said to incorporate information about the human bodies it is designed to fit, information the designer transmits to each garment by encoding that information in a pattern (or other formula) for producing every leotard in a particular batch. Whatever the full details of this "communication system," it seems evident that specific information about the positions to be occupied, the bodies to be covered, by the products of that system, has been incorporated in its reproductive "template."

Given the superb fit of the garment on the wearer, it may seem excessively cautious to avoid saying that the brand-new, unworn garment incorporates information about the wearer. And if such caution seems excessive, it appears simply ridiculous to be unwilling even to speak *as if* the garment incorporated such information, since that way of speaking can serve as a concise means of expressing what otherwise might require a long, tiresome description of the facts. Yet there is something to be said for tiresome alternatives to theoretically incoherent or misleading ways of speaking, however concise and otherwise convenient they happen to be; and while it might be an excusable distortion to say of a leotard presently being worn, that it possesses information about its wearer, the case is quite different for new, unworn leotards. The image that a small, shriveled piece of elastic material conveys of the body to which it will later conform is greatly impoverished; and extraterrestrial anthropologists could not reconstruct a reliable model of its potential wearers from the unworn garment itself. Nevertheless, the objection might be raised, even the leotard still in its package possesses information about its wearers: about their number of limbs, the approximate range of their dimensions, and so on. But although such information may have been essential to the design of the garment, that information is not obviously an image of its wearers that is conveyed (or transmitted) to generation after generation, batch after batch, of leotards. Indeed, information about the wearer may inform aspects of the design of the garment, without those aspects' directly representing features of the wearer. Thus, the openings for head, hands, and feet are not, as such, images of the wearer's neck, wrists, and ankles; rather, those openings can at most be said to correspond to and accommodate those anatomical features of the wearer. At the same time, those openings must accommodate the use of the garment as clothing for special purposes—accommodate the need to be able to put the clothing on, to prevent its legs from inching up the legs of its physically active wearer, and so on. And even those aspects (viz., legs, arms, and torso) of the design that can best be said to represent features of the wearer might better be described not as presenting an image of the wearer, but as affording the garment the *potential* to mold itself to, to yield an image of, the wearer. For the garment to do this, it is helpful, but not essential, that its parts correspond to parts of the wearer: A sufficiently elastic material could perfectly mold itself to the human form, without the benefit of preexisting parts that correspond to human bodily parts.

The lessons to be drawn about ethological concerns from these analogical remarks are these: It is the phenotype, not the genotype, that is adapted to its environment; and however well fitted that phenotype is to its surroundings, it does not follow that the genotype contains specific information about the environment or that the development of the phenotype is a process of decoding that information from the genotype. And even the phenotype may preserve a degree of elasticity enabling it to conform to environments different from those that originally applied selection pressure affecting the pattern of the genotype. Such a pattern cannot literally be said to encode information even about the environments originally affecting it, much less about the altered or alternate environments in which its phenotypes end up.

1.9 Rigidity as a Possible Criterion of Innateness

Given the aforementioned opposition between elasticity and innateness, it might seem reasonable to suggest that behavioral rigidity is indicative of instinct. A renowned advocate of this suggestion is M. Fabre. One of his persuasive examples concerns the behavior of the female Sphex wasp: After laying her eggs in a burrow, she searches for a cricket, stings it, and brings the paralyzed victim back to the burrow. Leaving the cricket outside, the wasp goes inside. She then returns to the cricket and drags it inside, where it is sealed in as food for her hatchlings. This order of events proves surprisingly inflexible: "Fabre discovered that, if he picked up the cricket while the female was inside the burrow and removed it a short distance away, the wasp would find the cricket and drop it at the threshold again before going back inside the burrow. Fabre moved the cricket away from the entrance to the burrow about forty times, and each time the wasp would drop the cricket at the entrance, never bypassing this link in the chain by dragging the cricket directly into the burrow." [48*]

The wasp's rigidity, its maintenance of an invariant order among the distinctive phases of behavior, seems compelling evidence for what Fabre called the "ignorance of instinct"—given which presumptively innate, "normally adaptive behavior becomes grotesquely maladaptive due to the insects' inability to respond to situations in a flexible manner." [49]

Perhaps some of the appeal of the idea that rigidity attests to innateness is attributable to an inchoate thought that such seemingly stupid behavior as the wasp's could not possibly be learned. Learning seems far too adaptive a process to account for the wasp's inflexibility in the face of dead crickets that won't stay put. If the wasp had actually learned her cricket-collecting behavior, she ought subsequently to be able to learn to modify that behavior.

But just how inane is the wasp's behavior? To answer fairly, one would need to know, among other things, whether any vital purpose is served by the wasp's leaving its victim outside, going inside, then retrieving the cricket. Perhaps this behavior is a matter of checking things out; and this practice may guard against other insects' secreting themselves in the burrow, where they, not the offspring, would benefit from the juicy cricket. If the female is thus merely being cautious, then her apparent failure to learn to behave otherwise is far from stupid: All the more caution is called for in the objectively suspicious circumstances created by Fabre.

But why suppose a mere wasp to be so astute? Isn't it far less anthropomorphic, more plausible to posit a special wisdom of instinct? Lorenz takes this line in relevantly similar cases, suggesting that some behavior is rigid for good (evolutionary) reason. Illustrating his point with reference to the likelihood of improving the performance of a "thoroughbred" sports car by having an amateur mechanic tinker with it, Lorenz suggests a general principle: "The more complicated an adapted process, the less chance there is that a random change will improve its adaptedness." [50] This principle is used to argue against the idea that there is some sort of "general and diffuse modifiability of phylogenetically adapted behavior mechanisms." [51] Those mechanisms are wisely rigid or, as he puts it, refractory to learning. If one countered with the suggestion that learning is not a merely random change, Lorenz's reply would be that when it isn't random, the change is more likely to be an outcome of "built-in learning mechanisms . . . phylogenetically programmed to perform just that function" [52*] than to be the result of diffuse modifiability. This reply appears to urge in favor of multiplying learning mechanisms beyond the necessity of any good evidence. And while we might agree with Lorenz that there is some reason to posit relatively function-specific learning mechanisms, the very generality of the terms he uses for them (e.g., "habituation") raises unanswered questions about just how function-specific these mechanisms really are.

Lorenz's—presumably more sophisticated—version of "the old and allegedly naive theory of an 'intercalation' of phylogenetically adapted and of individually modifiable behavior mechanisms" [53] suggests that innate behavior is rigid in the following sense: The phases of innate behavior are either wholly refractory to learning or subject only to specific, preestablished learning mechanisms. This theory might be applied to the Sphex wasp's behavior by depicting it as a complex pattern partly refractory to learning. The evolutionary wisdom of this intercalational rigidity might then conform to the intercalation theory, thus understood as a simple denial of the diffuse modifiability thesis.

Proponents of instinct might not be disturbed by the prospect of being able to apply the notion of intercalational rigidity so widely and might even welcome the further conclusion that so much behavior is at least partly innate. But however widely applicable this notion, the conclusion would be premature; for the possibility remains that the pattern has been acquired on the basis of experience. After all, a well-ingrained habit may also be thus rigid, yet it would certainly not qualify as instinctive.

Far from being criterial for innateness, intercalational rigidity seems, then, merely to provide inconclusive testimony. But what if this testimony were bolstered by other empirical findings, by positive results of the deprivation experiment? Such findings might be supposed to remove any lingering doubts, to rule out the possibility of showing any other way a whole species might come to exhibit certain rigid patterns of behavior. The deprivation technique clearly merits further scrutiny.

Chapter 2

Theoretically Questionable Experimental Support for Instinct

2.1 Problematic Applications of the Deprivation Experiment

The idea that information about the environment might be the innate component of certain behaviors is, I have tried to show, insupportable. One obvious alternative, though, is to suggest that some behaviors might themselves be instinctive. And while the deprivation technique should be deprived of Lorenz's sense of it, as a means of detecting an animal's innate information about its environment, what remains might still be thought a useful way to identify an animal's instinctive reactions, unconditioned responses, to various stimuli. Yet some major criticisms by D. S. Lehrman of the deprivation experiment were previously met by appealing to considerations about how information can be acquired; so if the nature-nurture debate is now to proceed without reference to information, the case against Lorenz's technique should be reopened.

Faced with the earlier question of whether a chick still in the egg might be capable of acquiring information about its posthatch environment, Lorenz plausibly enough answered no. But now that the underlying issue concerns the innateness of behavior, not of behavioral images of the environment, the question is, instead, whether an embryonic chick might be conditioned to respond (by pecking) to an exterior stimulus by being exposed to a comparable though causally distinct stimulus within the egg. To this question an affirmative answer seems reasonable: Even without any causal link between stimuli inside the egg and ones outside the egg, these two classes of stimuli could be functionally equivalent—that is, conditioning the embryonic chick to peck at the red spots on the interior of its eggshell might effectively condition that chick to peck, subsequent to hatching, at tiny red seeds.

The last answer may incline one to suppose that the contours of the concept of innateness are hopelessly blurred, but Lorenz has available the clarifying notion of what he once, facetiously, called the "innate schoolmarm," that is, a "phylogenetically programmed teaching machine."[1] The idea is that if the embryonic chick learns to peck at red seeds, then the chick must have been preprogrammed, genetically, to learn this from sources inside the egg. The notion of such a phylogenetically evolved "teaching mechanism" gives rise to a new category of behaviors—a category not simply of innate and conditioned behaviors (since that would also include any initially innate behaviors subsequently modified by experience), but a category of behaviors that are innate by virtue of being conditioned.

Lorenz regards this unusual category as a valuable addition to his ethological theory, an addition that allows him to acknowledge the possibility of embryonic conditioning without putting at risk any innateness claimed for behaviors thus conditioned. His point of insistence is that the channels of communication for environmental information are abundantly clear: Such information is not directly transmitted from outside the egg, from the environment itself, so the innate schoolmarm must be availing herself of genetically preprogrammed information instead.

If the nature-nurture question here concerns how the behavior is determined (rather than informed), then the nearest parallel to Lorenz's point is the claim that embryologically conditioned behavior should be regarded as genetically determined; but this claim is not obvious. Suppose that red spots appear on the inside shell of the egg only when the outside temperature is between certain specific limits and that, when these limits are exceeded, the absence of red spots precludes the chick's learning to peck at red seeds or any other small red target-stimuli.[2*] Has our imaginary chick learned its characteristic pecking from an innate schoolmarm or from the outside environment? Is this pecking innate or acquired? Finally, is there any reason to think that the deprivation experiment could be used to tease out the innate elements of this situation?

Since learning or conditioning prior to hatching is in question here, the deprivation experiment should be conducted on the embryonic chick. Given the details of the case as imagined, there is one obvious way to deprive the chick of the red spots on its shell's interior: Raise or lower the temperature beyond the prescribed limits. The danger is that we might also thereby change other processes crucial to the case: Who knows, for example, what effects extreme cold might have on an innate schoolmarm? But perhaps there is a less disruptive way of eliminating the spots: Suppose that very slight irradiation does this without otherwise affecting the chick or its present environment. By hypothesis, the chick, thus deprived, will not "learn" to peck at small red target-stimuli. On the assumption that embryological conditioning is requisite for manifestation of the behavior, a subsequent reintroduction of the shell spots will not immediately lead to their being pecked. The deprivation experiment will thus fail to establish the innateness of the behavior. Lorenz would presumably attribute the failure to a breakdown of normal development; but there is no "diffuse damage" to that development, and the hypothetical facts equally well support attributing the failure to inadequate nurturing instead.

Conducting the deprivation experiment on the hatchling instead of the embryo (which was exposed to shell spots) would lead to the conclusion that the pecking behavior is innate, since it would be exhibited immediately after the stimuli were (re)introduced into the environment. But all the hypothesized facts may not square with this conclusion: just what role did the temperature play in what transpired? The innate teaching mechanism used the red spots inside the shell to condition the embryo to peck at small red target-stimuli. Yet the environment outside the egg produced the spots; so, the embryo learned how to do something that is well adapted to, has survival value in, the external environment, and the embryo learned this behavior partly from that environment.

2.2 Tangled Environmental Influences upon the Development of Behavior

One defense of the deprivation experiment might have us distinguish between behavior learned in the egg or womb on the basis of the same external stimuli (e.g., a chick's mother's singing) with which the behavior will be coordinated and behavior learned by an embryo on the basis of different internal stimuli (e.g., red shell-spots), that is, the effects of other external factors (e.g., temperatures) that are causally independent of the external stimuli (e.g., red bugs or seeds) with which the behavior will be coordinated. Although the former sort of behavior does not deserve to be called innate, perhaps the latter sort does.

The assessment of the bird-song case seems trivially true: If the sounds of its mother's singing are the stimuli to which an embryonic chick has been exposed and from which it has learned certain recognitory behavior, then that behavior is, patently, a learned response to those same sound-stimuli. But why should any different assessment be made of the spotted-shell case? Given the possible reading of the phrase *same stimuli* as *functionally equivalent stimuli*, it might be said that the red spots within the egg are the same stimuli as the bugs or seeds with which the chick's pecking behavior outside the egg is coordinated. And if so, the pecking behavior might be adjudged no more deserving of the label *innate* than is the recognitory behavior.

What we mean by *same stimuli*, though, may instead have to do with causal connectedness, since the lack of such connectedness does seem to warrant our use of the term *innate* for behaviors otherwise learning-dependent upon the environment. Thus, the mother's song inside the egg is causally connected with the mother's song outside the egg, while the red spots inside the eggshell are not causally connected with (and hence not the same stimuli as) the red bugs or seeds outside the egg. Or are they? For all we know, the environmental factors that cause the spotted shell may also be factors essential to the presence, within that environment, of red bugs or seeds (e.g., if the temperature strays beyond that range, it may be too hot or too cold for other organisms to produce their tiny red offspring, the chick's favored diet).

Admittedly, the lines of causal influence are not fully comparable; for in the case of the mother's song, the exterior stimulus itself is *the cause of* the interior stimulus; whereas in the case of the red spots, the exterior stimuli are *caused by* the same factor that also, independently, causes the interior stimuli. But why should these differing lines of influence matter to the issue of innateness? According to Lorenz, the most important consideration is whether or not the embryonic chick has been exposed to the environmental exigencies to which the behavior is well adapted. So, notwithstanding the conspicuous lack of any direct line of causal influence from stimuli external to the egg to ones internal to it, Lorenz might agree that the chick species whose shell is sometimes spotted, sometimes not, has been exposed to the relevant environmental exigencies, to temperatures outside the egg that determine the availability of certain foodstuffs.

Moreover, a noteworthy feature of ordinary instruction is that the teacher need not always expose his or her students to (say) lions or tigers in order to teach them to respond appropriately to those beasts: for some instructional intents and purposes, a mere picture of these animals, combined with suitable commentary, will be quite sufficient. And while we have not gone into, have not imagined, the possible details of the instructional activity of an "innate schoolmarm," perhaps we ought to concede an analogous point for her: The innate schoolmarm need not expose the embryonic chick to bugs or seeds in order to teach it to respond appropriately to those target-stimuli—red spots inside the shell might suffice to introduce that chick to the right sort of stimulus-occasions for pecking after it hatches.

One may be tempted to dismiss out of hand any conclusion drawn from a far-fetched, purely hypothetical example. But the spotted-shell example is possible, and that's all I need to warrant my conclusion: The deprivation experiment is fully capable of wrongly identifying behaviors as innate.

2.3 Theoretical Insights Bearing on Inductive Failures to Identify Innate Behavior

Precise diagnosis of reasons why the deprivation experiment might fail to identify innate behaviors is as difficult as the situations in which that technique is used are complex. But a fuller appreciation of what is wrong with that technique might be had by considering theories (or theoretical outlooks) that call into question the whole idea of identifying innate behaviors. One such theoretical outlook is Lehrman's Schneirla-influenced "idea of continuity and interpenetration between the processes of growth and those of the influence of environment." [3] That idea, or something much akin to it, finds further expression in the following analogical remarks by the eminent Gestalt psychologist K. Koffka:

The intensity of an electrical current is proportional to the electromotive force, and to the conductivity of the system. . . . But it would be unreasonable to ask how much of this intensity is attributable to the electromotive force and how much to the conductivity of the system. . . . The fact is that the relation between a set of conditions and the process which takes place under these conditions is not as a rule such that we can divide the total process into a finite number of part-processes each of which will depend upon a certain part of the conditions.[4]

And applying this general rule to the specific case of human beings, Koffka concludes: "On account of his psycho-physical structure an individual possesses certain properties. These properties, together with his external social and physical situation, constitute the conditions of his behavior. . . . What we inherit, then, is not a repertory of particular reactions, but a set of internal conditions for response, which together with external conditions, physical and social, co-determine our behavior."[5]

Koffka's analogy may seem imperfect, since a mathematically sensible calculation of the proportionate influence of both electromotive force and conductivity upon intensity of electrical current could be made. More to the point of his analogy, however, is his contention that the determining conditions of a given process are not necessarily part-processes of the whole process: Thus, conductivity is not a part-process of the whole process of an electric current. The inherited response (or instinct) is for Koffka a bit of misplaced concreteness: a would-be part-process presumed to correspond to the endogenous conditions for response. The only real process is for him a whole process of behavior, which is co-determined by internal and external environmental conditions.

2.4 Instincts as Behavioral Impulses

Koffka proposes an almost mentalistic interpretation of the putatively innate internal conditions that co-determine various behaviors: "If we inherit anything specific, it is certain needs or stresses which pull us in certain directions. These needs or stresses result in responses which greatly vary with the external conditions under which the behavior occurs, yet remain constant in the direction of satisfying the needs which gave rise to them."[6] This account accords even better with the etymology of the term *instinct* than does the attempt to view behaviors themselves as instincts, for according to James Drever "instinct" is "in its original sense 'animal impulse' . . . hence a general term for natural or congenital impulse; erroneously used in modern times for forms of behavior rather than the underlying congenital impulse (behavior may be spoken of as *instinctive* . . . but not as 'an instinct'); more or less equivalent to innate or congenital 'drive' or 'need.' "[7*]

Unlike impulses, Koffka's "needs or stresses" are said to pull us, rather than to impel us, toward their satisfaction in behavior, but this difference seems of little moment: Does the positive pole push or the negative pole pull the electric current along its path? And to its credit, Koffka's account of these innate internal conditions does allow for a generally Darwinian view of their origins: "Needs" must be satisfied if the organism is to survive, so there is selection pressure favoring a species whose members are moved in the direction of behavior that satisfies those needs.

Moreover, Koffka's account does square, as it should, with Lehrman's biologically sound suggestion that environmental factors interpenetrate the processes of growth and development of individual organisms. Yet if it is to be adjudged a proper theory of instinct, Koffka's account should also satisfy a different, almost contrary desideratum: Much as a family tree might allow us to trace the origins of a given person back to one among many suspected ancestors, perhaps any acceptable theoretical account of instincts or instinctive (aspects of) behavior should enable us to trace their origins, to establish their distinctively hereditary determination. Indeed, one attraction of the myth that there are channels of communication for biologically inherited information about the environment may be that this myth does address a felt need to trace the very thing deemed innate back to the earliest phase of an organism's ontogenesis. Might there not be some other, philosophically more tolerable way to meet such a need than by endorsing that myth?

Koffka's own image of an electric current might be used to suggest a pertinent alternative: However indirect and complicated electrical connections might be, a discharge of current from one point to another implies that a line between those points does exist. Pursuing the analogy, let us consider what connections Koffka's account—or a mechanistic variant of it—might posit to enable us to establish hereditary lines of descent for endogenous co-determinants of behavior.

Need might be viewed as analogous to an electromotive force. Any set of external conditions that, in the capacity of stimulus, serves to release the behavioral response could be construed as a switching device that closes the electric circuit. Behavior could be equated, metaphorically, with mechanical operations of motors hooked up to the circuit. To complete this schematic diagram, something must be identified, by way of analogy, with the generating source of the electromotive force.

Gestalt psychologists tend to view need forces as "psychical field forces" and not to be too concerned, initially, about their physical bases. Thus, according to Kurt Lewin, "When the concept of energy is used . . . and when later those of force, of tension, of systems, and others are employed, the question may be left open as to whether or not one should ultimately go back to physical forces and energies." [8] This question is for Lewin one of whether psychology may ultimately be reduced to physics. But psychology is viewed by him as a part of biology, psychical processes as life processes; and so he assumes the motive force of need to be connected somehow with biological conditions of the organism. Thus a need for food is at once a biological condition and a motivating force. According to Koffka: "The need of food corresponds to a certain stress, or tension, and the activity that follows relieves tension."[9]

Yet impulsive need, which doesn't even exist in well-satiated organisms, is hardly the best candidate for innateness. A more promising alternative is the biological mechanism that, from time to time, generates such need—though that generator would also have to be part of a larger whole, the complete circuit that mobilizes the organism in ways deemed appropriate to the variety of need engendered.

Many details of this mechanistic variant of Koffka's view have yet to be worked out. So, for example, some means of shutting off the generator when conditions no longer call for need qua motive force are required. But some Gestaltists might question whether any such piecemeal approach, an accumulation of mechanistic details, could possibly capture the dynamic whole to be explained. John Dewey, though not a Gestaltist, makes a pertinently holistic point in his critique of the reflex-arc concept. He suggests inter alia that that concept fails to take account of the complex coordinations required between (what might better be called) "*stimulus activities* of the organism" (rather than, simply, "stimuli") and the "*motor-response activities*" appropriate to them: The former are said to "fix the problems" to which the latter are said to provide the solutions.[10] And while it could be posited that the activity-producing motor is, by "evolutionary design," already coordinated with environmental stimuli, that the resulting activity of (say) pecking is *pre*coordinated with certain visual target-stimuli, this views coordination as something achieved once and forever-after functioning properly; whereas Koffka, in basic agreement with Dewey, views it more plausibly as a process of continual adjustment. Thus, as Koffka puts it: "The situation which presents itself to the sense-organs, after a movement has taken place, determines the continuation of the movement."[11]

One could, however, devise mechanical means to accomplish the requisite task of continual coordination—say, by adding to the schematic account a feedback mechanism that readjusts the "motor" (and/or its mechanical attachments) to ever-changing environmental exigencies. Since these additions, like all the other components of this jerry-built variant of Koffka's account, would seem to involve distinct part-processes, the revised theory may depart from Gestalt orthodoxy; but that is no reason to reject the theory out of hand.

Might not some such theory help to vindicate the suggestion that certain internal conditions for behavior are innate? Those conditions (including impulse generators, activity motors, coordinating mechanisms, and assorted interconnections) have been rendered so morphological by the proposed theory that there seems no more reason to question their phylogenetic origins than those of any other bodily parts of the organism. And given neo-Darwinian theory's status as "the greatest unifying theory in biology," [12] the conjecture that there has been an evolutionary development of such determining conditions for behavior seems, if nothing else, a very good bet.

2.5 The Presumable Omnipresence of Environmental Determination

We have pursued our quarry, the innate component of behavior, ever inward— from external behavior alleged to involve innate information about the environment, to the impulses that help to determine that behavior, and now to the mechanisms that foster and channel those impulses. It might appear that we have finally cornered the beast but have found it more lively than we had anticipated. We might be reminded of Leibniz's famous quip about innate ideas: In response to the Lockean slogan that "there is nothing in the intellect but what comes from the senses," Leibniz adds, "*except the intellect itself.*" [13] So, too, however much we might wish to deny that nothing is innate in the organism, we must concede that the organism itself, its internal mechanisms and the organs that make it the organism it is, is "innate." Must we not?

We might instead harken back to the claim that environmental influences permeate the processes of ontogenetic development, and we could use that claim to suggest that the organism is what it is as much because of the environment in which it develops (including the environment within the egg or womb) as because of its genotype. This suggestion, which does indeed go quite far toward effectively eliminating the occasions for using the term *innate* as mutually exclusive of *environmentally determined,* could be said to preserve one final candidate for pure innateness: the DNA molecule (assuming, of course, that that molecule had not been subjected to any mutagenic environmental influences after fertilization). But there is something absurd about the idea that only the DNA molecule is truly innate, for that is all too much like saying that only of the letters of the alphabet can we be sure of the spelling: The DNA molecule is the ultimate (ontogenetic) basis of our genetic traits and is not itself a genetic trait.

But if our quest for innate elements has taken us too close to the core of hereditary determination, to its basis rather than to items determined on that basis, then the option remains of retracing our steps and reassessing the conception of innateness that has misled us in our prior search. The first step away from the center lands us back in an area of endogenous organic processes subject to environmental influences. Viewing our movement from center to periphery in embryological terms, from the DNA in the fertilized ovum to the mature functioning organism, it might be said that even the blastula (the first grouping of embryonic cells) is what it is because of its environment as well as its genetic makeup. So, if we are to make sense of the suggestion that some internal mechanisms associated with the organism's behavior are innate, we shall have to conceive of innateness in a way that circumvents the issue of environmental determination.

2.6 Relations between Learning and Orientation Mechanisms

It has seemed reasonable to assume that a theory of the evolutionary origins of mechanisms determining behaviors would do more to resolve the issue of their innateness than would a theory that simply explained behavioral functioning in terms of those mechanisms. But once we concede that those origins are mixed, a combination of genetic and environmental influences, then we must look elsewhere for a way to justify designating some behavior-related traits of the organism as innate.

It might be supposed that whenever any learning mechanism is requisite to the proper execution of certain behaviors, they may be said to be acquired, not innate. But consider behaviors that involve feedback mechanisms essential, say, for coordinating stimulus activities with response activities. Such feedback devices, especially if they enable the organism to cope increasingly well with the ever-changing demands of its environment, do seem to qualify as learning mechanisms; yet there might be some resistance to viewing the behaviors as acquired (rather than innate) merely because they involve these mechanisms. The sticking point would be that the mechanisms themselves seem, in Lorenz's words, "phylogenetically programmed to perform just [one, fairly specific] function." [14]

Alternatively, it might be suggested that behavior-producing mechanisms that operate independently of learning are the true standard-bearers of innateness and that the behaviors they produce are entitled, derivatively, to be called "innate." N. Tinbergen's account of the fly-catching behavior of frogs uses some ethologico-theoretic terms that further the cause of this suggestion:

> When a frog sees a fly it makes a sideways turn which results in its facing the fly fair and square. If the fly then comes within shooting distance, the frog will flip out its tongue. . . . The first movement is a taxis; it can be a mere turning towards the prey without any forward movement and in that event is a pure taxis. If, as is often the case, the prey keeps moving about, the taxis will be combined with locomotory patterns and a more or less complex oriented movement will be the result. The actual tongue-flipping, after having been released by the adequate visual stimulus situation, is not corrected. If the prey moves in time, the frog will miss it. The tongue movement is a true fixed pattern, though one of the simple types. [15]

Tinbergen contends that here, and in other cases, "taxis and fixed pattern are not linked simultaneously but are performed in succession." [16] And corresponding to these temporally distinct phases of the fly-catching behavior, there are presumed to be distinct behavior-producing mechanisms that operate (serially and thus) independently of one another.

On the assumption that learning might insinuate itself among the operations of an orientation mechanism (which governs a taxis) but not among those of a fixed-pattern-producing mechanism, the suggestion now under consideration would have it that the latter mechanism and the fixed pattern it produces are innate. But Lorenz would dispute this assumption, to insist on the innateness of the taxis, too. He says: "Among all the mechanisms serving the gain of immediately exploited information, those achieving orientation in space stand apart insofar as they determine not only at what moment but also in what direction a motor pattern is to be discharged. . . . They often produce 'learning' and frequently are its prerequisite. They are, however, totally independent of learning; they function even in the lowest unicellular organisms which do not learn." [17]

Of course this appeal to lowly organisms serves merely to show that orientation mechanisms can function independently of learning, not that they always do.[18*] Perhaps this lapse in reasoning attests to Lorenz's strong belief in the general possibility of learning-independent functioning. Since this belief is central to Lorenz's own thinking and does suggest a way of defending instinct without recourse to the deprivation experiment, the belief merits further consideration.

Lorenz's response to a contrary view is all too confident:

The assumption that learning "enters into" every phylogenetically adapted behavior mechanism is neither a logical necessity nor in any way supported by observational and experimental fact. Although practically every functional unit of behavior contains individually acquired information in the form of a stimulus-input which releases and directs reflexes, taxes, etc. . . . and thus determines the time and the place at and in which the behavior is performed, it is by no means a logical necessity that adaptive modification of behavioral mechanisms is invariably concerned in the survival function of these short-notice responses.[19]

But Lorenz's own contention, that many evolved behavior mechanisms function independently of learning, might be challenged in the selfsame way. His own case for that contention seems to rely heavily on the plausibility of ethological descriptions such as Tinbergen's; but there is no logical necessity to them, and their empirical support is, as we shall see, rather shaky.

Still, such descriptions do provide a vivid picture of animal behaviors as largely mechanical and instinctive. If, for example, the frog's fly catching is believed to consist of two phases, taxis and fixed pattern, both of which are produced by the learning-independent operations of certain mechanisms, then the frog takes on the guise of something like a computer-controlled, turret-mounted piece of

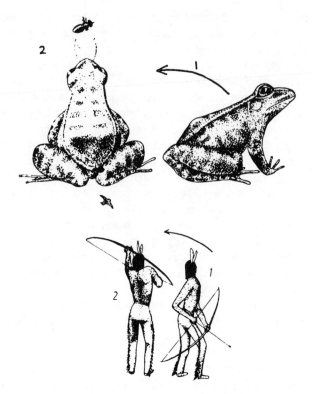

Figure 1. Aiming and subsequent shooting in frog and in primitive man, from Niko Tinbergen, *The Study of Instinct* (Oxford: Oxford University Press, 1951). Copyright © 1951 by Oxford University Press. Used by permission.

artillery: First the turret turns, automatically, until the target is in line with the trajectory of the gun; then the device fires, again automatically. This construal is easier to swallow in the case of frogs than it is when applied to human beings; but Tinbergen himself, without comment, juxtaposes a line drawing of the frog and another of a Native American holding a bow and arrow (see Figure 1). Tinbergen is surely right if he thinks that this picture speaks for itself. Our suspicions ought to be aroused by these strange impressions.

2.7 Efforts to Individuate Simple Movements

Tinbergen and Lorenz seem to base their claims about autonomously functioning, behavior-producing mechanisms on little more than the possibility of identifying separate and distinct movements that make up an organism's complex behaviors (or movements). Thus, Tinbergen assumes that individual mechanisms correspond to each of the frog's distinguishable movements. But this immediately raises the question of how these movements are supposed to be correctly counted. Tinbergen seems to regard "mere turning towards the prey without any forward motion" as one simple movement; and so, no doubt, moving sideways and forward would be counted as a complex movement. But on what basis are the two component movements to be individuated? Direction alone would not provide a satisfactory differentia, since sideways-forward (or oblique) motion could be viewed as movement in only one direction rather than two. Direction and temporal succession would serve to pick out two movements, provided that the frog did move serially and not simultaneously. But then smooth movement simultaneously sideways and forward really ought to qualify as one movement rather than two, whereas Tinbergen wishes to assert that two distinct mechanisms always control sideways and forward movement.

An inventory of the muscles used by the frog in producing various movements might be thought to provide the criterion Tinbergen needs, but just what sort of criterion would this be? We might suggest that the fewer the muscles involved in producing a given movement, the simpler the movement would be—though, for all we know, the frog may use fewer muscles in moving its head simultaneously sideways and forward than in "merely" moving (indeed, rotating) it sideways. And our ability to compensate for an injury to one muscle by producing the same movement with other muscles suggests problems for a pat equation of movements and muscles in use. A defender of this equation might observe that, since muscles in use themselves move, it is tautologically true that using different muscles means making different movements. But if the defender of the criterion goes that far to defend it, let us go even farther in the same direction to attack it. Twice flexing the same muscle involves different muscle fibers to varying degrees, so perhaps no two movements of the organism are truly identical: We cannot step into the same river twice not only because the river changes, as Heraclitus observed, but also because our movements are never twice precisely the same step.

It should also be observed that simple muscular contractions are not bound to be additively implicated in complex bodily movements: The (range of) contractions involved in two simple movements—say, sideways and forward—(is) are not likely to be identical with those involved in a continuous complex movement— say, smooth movement in an oblique direction. For one thing, the continuous complex movement would not necessarily require the type of constantly varying direction involved in rotational sideways movement: The complex movement might be unidirectional over its entire duration. For another thing, moving one's limb (e.g., an arm) straight forward and up from the body may require the use of one muscle; raising it sideways and up may require the use of another muscle; and raising it at an angle halfway between the other two movements may require the use of yet a third muscle but not involve either of the other two muscles. These points are, to be sure, anatomically conjectural and may not be correct; but they do suggest some possibilities that should be ruled out, by empirical investigation, before settling upon a muscle-inventory criterion for movements.

2.8 A Phenomenological Defense of Animal Intelligence

Lorenz and Tinbergen believe that it is possible, experimentally, to tease apart the simple components of complex oriented movement. Their joint efforts to do so presuppose that such behavior is blindly mechanical rather than intelligent, and Lorenz even denies any real difference. He contends that the orientation "mechanisms imparting instantaneous temporal and spatial information are, in their more highly developed forms, largely identical with what is called intelligence in common parlance." [20] Selected for detailed study are the greylag goose's efforts to retrieve an egg that rolls out of her nest: She balances and pulls the egg back with her bill. Sometimes, when the egg slips away, the balancing movements stop but the pulling motions continue. As Tinbergen describes it: "The egg-rolling movement does not always break off, but it may be completed, very much as if it were a vacuum activity. If this happens, the sideways balancing movements are absent. This indicates that the balancing movements are dependent on continuous stimulation from the egg, probably of a tactile nature, while the other component, a movement in the median plane, is not dependent on continuous stimulation but, once released, runs its full course." [21]

In order to challenge the Lorenz-Tinbergen account, to defend the goose against their automatonlike construal of her, I will try to put myself in her place and think like a (moderately intelligent) goose. My proposed procedure, though out of zoological fashion, is not unprecedented: Jakob von Uexküll, in his engaging monograph, *A Stroll through the Worlds of Animals and Men*, tries to describe what he calls the *Umwelt* (i.e., the phenomenal world) of creatures as alien to us as the eyeless, bloodsucking tick; and his motives for doing so are not unlike my own. He says:

> The mechanists have pieced together the sensory and motor organs of animals, like so many parts of a machine, ignoring their real functions of perceiving and acting, and have gone on to mechanize man himself. . . . We no longer regard animals as mere machines, but as subjects whose essential activity consists of perceiving and acting. We thus unlock the gates that lead to other realms, for all that a subject perceives becomes his perceptual world and all that he does, his effector world. Perceptual and effector worlds together form a closed unit, the *Umwelt*. These different worlds . . . present to all nature lovers new lands of such wealth and beauty that a walk through them is well worth while, even though they unfold not to the physical but only to the spiritual eye. [22]

One preliminary hypothesis, which will help me to discern some good sense in the goose's behavior, is that her lower bill is significantly more sensitive toward its outer sides than along its middle (alternatively, that the bill is more flexible away from its center, thereby permitting tactile pressure closer to its sides to be better transmitted and felt at other sensitive locations nearby).[23*] Here, then, is a not-so-silly goose's *Umwelt* during egg retrieval: Seeing the egg before her, where it has rolled from her nest, the goose, still inside the nest, stretches forward and places her bill over and behind the egg, from which position she can use her bill to pull the egg back toward the nest. She begins pulling, and whenever she feels the egg veering off to one side or the other, because she feels it on that side of her bill, she moves slightly to that side and resumes pulling. This activity typically continues until the egg is back in the nest. Sometimes, though, the egg rolls too far to one side to be balanced effectively, and soon the egg is out of reach altogether. The goose either does not notice that the egg has gotten out of reach, because she is too absorbed in the activity, or cannot see the egg, for anatomical reasons (just as we couldn't see small objects under our chins), or, if she does notice the egg rolling away, does not know whether it is not yet another egg. Not feeling the egg, which is fairly light and rolls easily, is a sign, for the goose, that everything is going well, that there is no need to engage in sideways-balancing movements. Since she assumes that the egg is still under her bill, the goose continues to pull all the way back to the nest.

Now if we construe intelligence as doing what is called for, taking appropriate steps, given what one knows about the environment, in order to achieve one's goal; then this behavior, far from being a blind mechanical reaction to a releasing stimulus and an equally blind nonreaction to a lack of other stimuli, is genuinely intelligent. One might wonder why, if she is truly intelligent, the goose does not make a few slightly sideways moves in order to assure herself that she is still in contact with the egg; but, given the tricky business of balancing and rolling an egg-shaped object, this really would not be very smart: It would be as if a tightwire walker were to look down at the wire to make sure that her apparently successful efforts were really transporting her over an abyss, as if an impatient fisherman were to jerk at his line in order to make sure that a soft-mouthed fish was still attached to the hook. If indeed the goose feels nothing, then perhaps she should have her doubts (or whatever it is that geese can have). But so, too, the fisherman should sometimes have his doubts and yet, if he is a smart fisherman, still cautiously play his line as if he were sure that a fish remained on the hook.

Lorenz and Tinbergen attempt to distinguish experimentally between two components of the goose's egg-retrieval behavior: the sideways-balancing "directed" reaction, said to depend on continuous stimulation, and the backward-rolling "released" reaction, said to run its full course after an initial stimulation. In the absence of clear criteria for individuating movements, this distinction seems conceptually problematic: One could just as well construe the sideways balancing as a complex set of individual reactions, each of which is fully released by a single

Figure 2. Greylag goose retrieving egg and attempting to retrieve giant egg, from Niko Tinbergen, *The Study of Instinct* (Oxford: Oxford University Press, 1951). Copyright © 1951 by Oxford University Press. Used by permission.

tactile stimulus. The observational evidence, moreover, does not clearly support such a distinction: Tinbergen tells us that, when an egg slips away, "the egg-rolling movement does not always break off";[24] but this suggests that sometimes the movement does then stop, contrary to the released-reaction hypothesis. And the experimental findings are surely inconclusive. According to Tinbergen: "The stereotyped nature of [the backward-pulling] component was especially obvious when the bird was offered an egg that was much too large. It could not adapt the movement to the abnormal size and got stuck half-way when the egg was pressed between the egg and the breast."[25]

The conclusion that this behavior is stereotyped is supposed to reflect badly on the goose's intellect and/or her capacity to learn: She just cannot seem to adapt. But putting myself (with cognitive capacities intact) in her place, I do not imagine I could do much better. Judging from the way the egg facsimile is firmly lodged between her lower bill and breast in Tinbergen's illustration (see Figure 2), continued pulling movement seems mechanically impossible; and the goose can hardly be faulted for failing to figure out how to do the impossible.

Tinbergen also reports another experimental manipulation of the goose's behavior: "The balancing movements could further be eliminated by presenting the goose with a cylinder. This model, when offered on an entirely flat and smooth surface, made no sideways movement. The result was a straight egg-rolling movement without any sideward deviations of the head."[26]

I have no quarrel with the conclusion that egg-rolling movement does take place in the absence of egg-balancing movement, but I do question Tinbergen's theoretical inference that the two movements are produced by two independently functioning internal mechanisms, each subject to a different sort of stimulus-control. The goose is being construed as a relatively simple automaton. But if we view the goose as somewhat intelligent, her behavior is not surprising: The point of her sideways-balancing movements is to balance unstable objects. Since the cylinder is stable, there is no reason for her to engage in such movements. Moreover, the absence of those movements does not establish that the goose is not engaged in balancing—any more, say, than a man's remaining motionless shows him not to be balancing an inverted broom on his nose. The only aspect of her behavior that might raise questions about the goose's intelligence is her being taken in by such a poor facsimile of an egg. But is there any reason to suppose that she takes the cylinder to be an egg? Maybe the model just looks like something pleasant to sit on, or maybe toying with it seems a happy diversion from her interminable brooding.

The goose's egg-retrieval behavior might best be described, in Deweyan fashion, as a complex coordination of activities only functionally distinguishable from one another. And the same might be said for the frog's fly catching, even though the successive rather than simultaneous performance of the "taxis" and "fixed pattern" components makes it tempting to separate them. Each of these components figures essentially in the overall coordination of an active organism attempting to catch some food. The frog's functionally distinguishable activities are lining up and lashing out at the fly. The frog who does not first move sideways may nonetheless have lined up a fly properly; and the frog who does not line up a fly cannot be said to lash out at one. The invariant order of taxis before fixed pattern poses no problem that needs to be solved by postulating two successively operating behavior-producing mechanisms. Given what the frog is doing and the range of movements available for the purpose, it stands to reason that, whenever both sorts of moves are made, tongue flipping takes place after sideways turning.

2.9 Manifest versus Scientific Images of Intelligent Behaviors

The point of the above critical remarks about mechanistic conceptions of animal behaviors is not so much to insist that an alternative view of animals as actively involved with their environments is unquestionably superior, but to observe, more modestly, that this alternative is no worse than its rival. Still, it might be urged that the Lorenz-Tinbergen view does have more scientific standing than my alternative. Wilfrid Sellars has contrasted what he calls the "manifest image" and the "scientific image" of man-in-the-world, where the former "is to be construed as a sophistication and refinement of the image in terms of which man first came to be aware of himself as man-in-the-world; in short, came to be man" [27] and the latter is "the image derived from the fruits of postulational theory construction." [28] My alternative could be taken as an attempt to extend the manifest image to frogs and geese, to see them as somewhat intelligent fellow-beings-in-the-world; while the rival position, with its posited mechanisms, would seem to fit squarely within a scientific image. But there is more to science than "postulational theory construction." Such theoreticity is, after all, also characteristic of mythology and theology; and insisting upon the scientific status of the mechanistic view of the ethologists is a bit like saying of doggerel verse that at least it is poetry.

The modern preference for views with higher (as against lower) "grades of theoreticity" [29*] is akin to the ancient philosophic idea that what is real is not what is apparent to our senses. But while only the theories of our scientific image might truly reach reality's depths, that is no reason to dismiss our contrastively superficial manifest image as mere appearance. Only empirically well-supported theories could thus enhance our grasp of reality; and since support for them requires connections, if only via other scientific constructions, to the world as manifest to us, the idea that we should dispense with ordinary (manifest image) conceptions in favor of scientific ones smacks of sawing off a limb on which we stand. Such folly might be avoided by first constructing another means of support—say, by reconceiving the manifest image in terms of the very theories supported by it. But this procedure has its own risks—for example, it might serve to beg questions about how well supported those theories actually are, and it might be used to raise unwarranted doubts about the reality of palpable experiences that do not happen to lend themselves to scientific reconceptualization.

Now maybe we should try to reconcile discrepant images of reality, and no doubt reconceptualization is a handy tool for the purpose. But instead of just using this tool to transform the ordinary, we might sometimes reasonably decide (as well) to reconceive scientific theories in terms more agreeable to our manifest image. Of course we are not yet faced with the issue of reconciliation, since there isn't any well-substantiated scientific theory of animal behavior available to us; but if the mechanistic view anticipates a future need for reconciliation, if that view is a preparatory quasi-scientific reworking of the manifest image, then my alternative view hopes, rather, to forestall that need, by encouraging the search for a scientific theory consonant with a plausible extension of the manifest image.

But is my extension any more plausible than the mechanistic reworking? Perhaps the manifest image is more amenable to conceptual revision than I think. Suppose someone invents a frog-robot with motor mechanisms that can produce movements akin to the sideways, forward, and tongue-flipping movements of an actual frog. The mechanisms are operated, via a remote-control device, by a human operator staring at a television monitor hooked up to miniature cameras on the robot, in place of eyes. Once the operator has mastered the use of the controls, the sideways and forward movements might, despite the fact that they were produced by discrete mechanisms, be smoothly coordinated into seemingly undifferentiable wholes. Does this hypothetical case not suggest that a manifest image of behavior in general might easily be recast in mechanistic terms?

Let us not forget the role played in the example by centralized—albeit re-mote—intelligent control. Without such control it is very doubtful that compa-rable results could be achieved. But the real frog and goose are, according to Lorenz and Tinbergen, automata controlled by external stimuli that either fully release or continually direct the operation of behavior-producing mechanisms. Tinbergen's larger theory of instinctive behavior does involve a hierarchical ar-rangement of the individual mechanisms, though this is not intended to achieve centralized control, only to ensure the proper temporal sequence of events. The well-nigh inevitably jerky movements of beasts who fit this theory would betray their inner clockwork, show them to be mere automata. A clever designer might arrange mechanical linkages among the separate mechanisms, linkages designed to do no more than prevent jerky movements; but nature is not likely to be that designer: What survival value would those linkages have? Smoothly coordinated movement is advantageous, but the chief value of the smoothness per se is tied to the efficiency of movements in pursuit of vital goals. To ensure such effi-ciency almost demands centralized planning and control of these movements, and peripheral linkage devices would likely buy increased smoothness at the cost of efficiency.

While not conclusive, these considerations begin to favor my extension of the manifest image over Tinbergen's mechanistic, instinct-theoretic revision of it. Perhaps, though, the chief point of disagreement is not that Tinbergen's account is excessively mechanical, just that it credits animals with too little intelligence, be it mechanical or not. Yet if centralized planning and control are what his account lacks, they might be supplied without rejecting all the other features of his theory: That proverbial black box, the brain, might mechanically govern an animal's movement-producing mechanisms in a smoothly coordinated manner. The reality of manifestly intelligent behaviors might then be very similar to *some* ethological image of them. However, such increased emphasis on the role of intelligence in behavioral functioning might argue against the pretensions of instinct.

2.10 Intelligence as an Alternative to Instinct and to Learning

Intelligence and *instinct* are major explanatory rivals. Rather than view other members of the animal kingdom as intelligent, people often dismiss the behavior of other species as *merely* instinctive. But Montaigne argues: "There is no appar-ent reason to judge that the beasts do by natural and obligatory instinct the same things that we do by our choice and cleverness. We must infer from like results like faculties, and consequently confess that this same reason, this same method that we have for working is also that of the animals." [30]*

Following Montaigne's lead, we might infer from the smoothly coordinated movement of animals that their behavior, like ours, results from a faculty of intel-lect. And, insofar as we choose to oppose the operations of intellect to those of instinct, this affords added grounds for disputing the innate underpinnings of some animal behavior. Viewing animal behavior from the long-range perspective of phylogeny, it is easy to see how intelligence may, on occasion, be a more adaptive faculty than learning is. Thus, in a highly changeable environment, the lessons of the past might be impediments to coping with the future. And even in the case of complex coordinations within relatively stable environments, the value of learning might not be very great: Some flies will get away if the frog fails to anticipate their avoidance maneuvers, but learned anticipations would be efficacious only if flies frequently used almost identical patterns of avoidance.

Intelligence and learning are often combined in mutually supportive ways, but is there any conceptual inevitability to this symbiosis? Suppose that, in light of failed efforts, an organism uses its allotted intelligence to readjust its movements, to coordinate them with each other and with the environmental objects upon which they are directed. Since the readjustment need not be permanent, it might be argued that this exercise of intelligence need not involve learning. But doesn't there have to be some earlier learning to inform any genuinely intelligent efforts at readjustment? The notion of blind readjustment, although not conceptually incoherent, does conjure up an idea of nonintelligent, wholly mechanical functioning. Intelligent readjustment would seem to require a sense of what one is trying to do (with or without any real know-how). Of course this sense might be alleged to be innately provided, but that would seem to confuse an original capability, a facility for sensing, with one of its specific developments, an occurrent sense that something different needs doing. This occurrent sense presupposes something learned: some knowledge that one's past efforts have failed—knowledge that is one's basis for attempting something different the next time. Such knowledge could not be innately possessed, except in the thoroughly unlikely circumstance of a preestablished harmony between the knowing organism and its environment. So it seems that even in one of its most rudimentary modes of functioning, a mode that could in theory be simulated by means of what Lorenz calls an orientation mechanism, the intelligent control of behavior involves learning. I shall hazard the thus informed conjecture that learning is indeed integral to all genuinely intelligent behavior.

If this hypothesis is correct, then intelligence does not stand alone in its explanatory opposition to instinct. But this does not mean that intelligence is somehow a less worthy opponent than it would be if it did; for intelligence may, despite the help at hand, be able to win the fight unaided. In truth, it is a mistake to suppose that a subsidiary conflict between the processes of learning and "pure" maturation might help to decide the outcome of the larger dispute; for although the latter processes are sometimes in direct competition for involvement in the intelligent control of behavior, their coveted roles are both merely in the service of intelligence. So even if, contrary to what I have hypothesized, some intelligent behaviors were to involve no learning, only maturation, those behaviors would arguably not qualify as instincts. Still, one may again be reminded, less obliquely than before, of Leibniz's rebuke of Locke; for it could be insisted, with great plausibility, that the intellect itself is innate, that its hegemony over an organism's behavior is instinctive. This philosophical move merits further study within a biological context.

2.11 Computer Simulations and the Faculty of Intellect

Native endowment with intelligence is said by some believers in divine creation to be a gift of God instead of a gradual outcome of evolution; other believers hold that mechanisms of evolution are the means by which the gift is given. But despite their disagreement both religious viewpoints seem compatible with the idea that the gift is embodied in biological processes. Both viewpoints also seem compatible with the further suggestion that these biological processes are functionally equivalent to the operations of a highly sophisticated computer. Now this image of intelligence as (some of the functions of) a computer in the cranial cavity has clear advantages for someone who accepts an evolutionary perspective on biological phenomena: It permits mention of some protoplasmic material that might have been molded, by evolutionary forces, into its present convoluted state. If intelligence is instead seen as the attribute of a purely mental substance, then all the problems that confronted Descartes when he tried to explain how a mental substance might interact with a physical one come back to haunt us as we try to explain the biological genesis of intelligence.

But could a computer account for the types of intelligently controlled behavior that animals, even frogs and geese, produce? This question is sometimes understood to ask whether a computer could be devised to mimic exactly this behavior. Assuming that the frog-robot mentioned earlier is a reasonable facsimile of a real frog, the question might be concretized by asking whether a computer could take over the controls and produce behavior indistinguishable from either the food-catching behavior of a real frog or the accurate simulation of that behavior produced when the frog-robot is remotely controlled by a skilled, intelligent operator. (For aesthetic parity, let us say that the computer, corresponding to a frog's brain, is located inside the frog-robot's head.)

Without troubling to give any real specification (beyond purely verbal allusions to governors, coordinating mechanisms, etc.) of how the feat might be accomplished, let us suppose that the computer simulation is a resounding success. Since Montaigne's principle—namely, like behavioral results imply like faculties—was earlier used to warrant the inference that a frog has the faculty of intellect, it now seems only fair to rely on that principle to infer that the frog-robot, too, has this faculty.

But the import of the principle is rather unclear. How much alike are the "like faculties" inferred on its authority? If a vinyl record and a cassette tape of the same bird's song produce qualitatively indistinguishable results when played through the same amplification system and speakers, then the record and the cassette systems for sound reproduction would, given Montaigne's principle, be said to be like faculties. But where does their likeness begin and end? Going back to the source of the sound, we find complete identity—the same bird singing on the same occasion. The identity might extend to the sound pickup (one microphone is used) and the original recording (used as a source for both the record and the cassette). The processes of producing the final record and cassette would differ, as would the record turntable and cassette deck used to play them. The electrical impulses from the playback devices to the amplification system might also vary (e.g., in strength), but the amplifier might automatically compensate for the differences. Finally, the signals from the amplifier to the speakers might be identical (so far as we could tell). If these are the facts of the case, Montaigne's principle should not allow us to infer anything at variance with them. To avoid triviality, Montaigne must not simply define "like faculties" as ones that produce like results. His principle might substantively suggest, of the present case, that *something* similar must be producing the similar audible vibrations of the loudspeakers; yet unless we have at our disposal independently obtained information that these speakers are operated by electrical impulses, we could not even identify that similar something as a series of such impulses.

Suppose that our frog-robot is a dead frog whose brain has been replaced by a miniature computer and that electrical impulses from it control the movement of this fleshy robot just as they do when they originate from the brain of a live frog. Knowing all this, with or without Montaigne's principle, we can trace the likeness of the behavior-producing faculties back to the computer and the brain, respectively, but no farther. For all we know, computers and brains produce their impulses by very different means.

But is there not something about the behaviors themselves, their manifest intelligence, that permits us to go back much farther, all the way to their sources within the brain and its computer counterpart? Consider again the case of the recorded bird song: Does the pattern of sounds at one end not lead us, correctly, to the conclusion that the source of both recordings must, despite the differences in intervening processes, be the same? The conclusion is correct, but only by hypothesis; the argument is invalid. Suppose that the one recording is of a robin and that the other is of a mimic-thrush, the mockingbird. Despite the qualitatively indistinguishable bird-song phrases produced by the two recording processes, the source of the sounds is not the same. In fact, qualitatively indistinguishable recordings could even have been produced from easily distinguished source sounds—if, for example, the recording processes were low fidelity.

Yet perhaps the sound-reproduction analogy does not get at the heart of what is special about intelligent behaviors; and perhaps Montaigne's principle does hold true of them. What is special about them is that they have been devised as appropriate means to given ends. Are we entitled, then, to argue backward from the intelligent character of two behaviors, the live frog's and the frog-robot's, to like faculties that have produced them? The following case might occasion some doubts: One animal devises intelligent behavior to achieve its end; another animal blindly mimics the first and thoughtlessly achieves for itself the same valuable end, some food. Were we unaware of the mimicry, the two behaviors might strike us as indistinguishably intelligent outcomes of the same faculty. Mimicry can of course be intelligent behavior, whether devised for its own sake or for the sake of attaining the same end sought by the animal whose behavior is imitated; but blind mimicry may not be very intelligent at all. If it is not, then there is no reason to suppose it and the undeniably intelligent behavior that was its model to be outcomes of like faculties.

But far from contradicting Montaigne's principle, this last claim might seem, contrapositively, to endorse it. Nonetheless, the above case does prepare the way for some telling objections—yet to be raised—to the principle.

2.12 The Process and the Products of Intelligence

In order to invent a frog-robot that employs autonomously functioning mechanisms to simulate the behavior of a live frog, much intelligent forethought would be required. But once the task is finished, there no longer needs to be any faculty of intellect to control the robot's behavior. The sure indication of an intellectual faculty is not the devised but the devising, not the solution to a problem but the problem solving that terminates in that solution. Artifical intelligence theorists are wont to seek rules that, when programmed into a computer, yield solutions to well-defined problems. Having found such rules, some theorists feel entitled to claim that the problem-solving process has itself been simulated. This simulation, though, is really just a solution masquerading as a process of problem solving. Indeed, no temporal component is even requisite to the alleged process: A written version will suffice not merely as a description of the alleged process (as, say, written calculations using physical laws may describe or predict the behavior of the stars), but as the thing itself, the simulation. No real process is atemporal.

AI theorists might defend their artifact by pointing out that when it is programmed into a computer, their product becomes a process. And they might counter the further rebuke that no process ever could be timeless by denying that any timeless (e.g., merely written) incarnation is a proper instance of their simulation. However, this denial would be antithetical to the spirit of AI theorizing, which emphasizes the importance of abstract structural identity and downplays the need for specific kinds of realizations for any given structure. Functionalism, the prime example of such theorizing, does use the notion of sequential relations to characterize structural identity, but there is no obvious reason why these relations need to be temporal in every realization, why (say) in a written incarnation "*A* . . . *B*" could not be construed as *B*'s being sequentially related to *A*.

Quite apart from these functionalistic scruples, there is something suspicious about according a privileged status to fully specified solutions that just happen to, or are made to, unfold over time. In a genuine process of problem solving, one wants to say, success is not so completely foreordained. But in further defense of their alleged simulations of that process, AI theorists might here be expected to invoke the distinction between *algorithmic* rules, which afford infallible guidance toward a solution, and *heuristic* rules, which tend to result in a solution, although their "results are variable and success is seldom guaranteed."[31*] The defense would be that although algorithmic rules may provide little more than a solution, the same cannot be said of heuristic rules. To which an unsympathetic critic might respond, "That's true—heuristic rules don't *even* do that." Attempted simulations in which heuristic rules are automatically followed still replace the process of intelligence with its product, a clever device not of the computer's own devising.

All this may seem unfair to computers, so it should be remarked that much the same point applies to us: once we devise algorithmic or heuristic rules to solve our problems, genuinely intelligent functioning can come to an end. When next those rules are used, intelligence can effectively be supplanted by habit, which may or may not be blindly obeyed.[32*] Indeed, even the first time we use rules of our own devising, they may, conceivably, be deployed in a mechanical way thoroughly lacking in the exercise of intellect. Of course many occasions for the use of rules pose further problems about how to apply those rules, thereby demanding intelligence once again; but we should not suppose that there is always a problem about how to apply a rule, for then we would be in danger of an infinite regress: we might need to come up with an unending series of rules for applying rules. Intelligence may overlap, but should not be identified, with rule-governed behavior. Intelligence is, among other things, rule-generating behavior—except, that is, when the rule generating is achieved by mechanically applying other rules, whether algorithmic or heuristic.

An intellect-simulating computer would need to generate problem-solving rules in response to novel environmental exigencies. But why think that such a device would correspond to biological reality? Montaigne's principle notwithstanding, we would have no particularly good reason to suppose that a frog that produced the same intelligent solutions to problems of food catching as a robot-frog did would have generated those solutions by means of a faculty like the frog-robot's rule-generating device. That supposition would be rendered plausible if we knew, further, that the brain functioned like a computer; but we cannot be said to know this on the basis of a few facts about "how individual cells—the neurons—generate and transmit the electrical impulses that form the basic code element of our internal communication system."[33] We may no more conclude that the brain functions, on the whole, like a computer from the bare fact that (some of) the brain's parts generate and transmit electricity than we may conclude, from the same fact, that an old-fashioned record player also functions, overall, like a computer. Even if the "wiring diagram" of connections among neurons corresponded exactly to the wiring diagram of an incredibly complex computer—one with a hundred billion "switches" and a hundred trillion connections among them—it still would not follow that the brain functioned like a computer: Various "transmitter chemicals" (serotonin, gamma-aninobutyric acid, etc.[34*]) might influence the functioning of the brain in ways wholly unforeseeable on the basis of the comparison with a computer. The brain, awash in potent chemical agents, may be only superficially similar to a computer; it may in fact function very differently.

2.13 The Role of Problem Grasping in Genuinely Intelligent Functioning

Even if the brain could be assumed to function like a computer, AI theories of intelligent behavior would be difficult to reconcile with the manifest image of such behavior; since any truly intelligent, problem-solving behavior does appear to require the problem solver's grasping of a problem as a problem. Of course this requirement might be disputed by appealing to our own case: Do we not sometimes work out and even solve problems we do not grasp fully, if at all? But utter blindness to the problems we solve is possible, I think, only when our solutions are accidentally rather than intelligently devised. It is possible to work on "problems" whose solutions are the result of our mechanical application of certain rules; but these are more in the nature of exercises than genuine problems, and working on them demands no real intelligence. It is also possible and quite common not to recognize the scope and limits of a problem with which we intelligently grapple, but this is importantly different from failing to grasp a problem as a problem. This is more a matter of failing to appreciate the significance of a problem we do grasp—indeed, we cannot properly be said to fail thus to appreciate a problem unless we do grasp it as a problem.

All well and good for human beings, one might think, but what about frogs and geese? If frogs do behave intelligently in coordinating their food-catching movements, then they too should be viewed as grasping a problem, the problem of catching food. The frog may not know that fly swallowing will diminish its hunger pangs, but perhaps the frog thinks (roughly), "Want that fly—to eat!" This is not meant as a translation of the frog's cry, "Ribbit": The frog is not merely without language per se, but even without conceptual pegs on which to hang the terms *want, fly,* and *eat.* Unless it is innately equipped with some notion that moving targets afford appetizing morsels, the frog will initially lack any sense of the larger significance of the problem of fly catching. Later on, the frog might possibly develop some such sense, which would doubtless fall short of an appreciation of the nutritive value of food. From the start, however, the frog will have a diffuse interest in attaining the goal of obtaining the fly. This sense of a problem to be addressed is a precondition for conceding to the frog a measure of intelligence—a measure to be deployed (e.g., by coordinating movements) in the service of fly catching.

Now it might be suggested that the frog's initial interest is an innate impulse. To counter this suggestion, I might mention the alternative possibility that intelligence itself is what finds things interesting, grasps things as problematic, and so on; and then I might observe that there is an indeterminacy between the former suggestion and the latter possibility, that simple experimental means could not readily decide which of the two hypotheses has more merit. A well-confirmed theory of intellectual functioning might urge in favor of one over the other, but there is no such theory to which we can presently appeal.

Both hypotheses, native interests and an intellectual faculty that finds things interesting, are compatible with genuine intelligence; both allow for problem grasping, relative to which intelligence may play a problem-solving role. On either hypothesis, the interest is a concern not merely for fast-flying objects, for flies, but for gaining possession of those objects, for *catching* flies. It should be inquired whether an AI theory of intellectual functioning is compatible with the existence of such interests. Were they ready-made graspings of problems as problems, then perhaps these interests might be thus compatible. But how could anything more than a potential for grasping a problem precede exposure to something that poses a problem? Even native interests could not be inborn as actual problems for the frog. An AI theory might simulate fly-catching interests by having a frog-robot prepared to go after flies *as if* it always had the actual problem rather than a potential for grasping it. Like some sophisticated artillery that, in response to specific radar signals, aims and fires automatically, the frog-robot never really grasps a problem, never wants to catch and eat a fly. Genuinely intelligent frogs do.

AI theorists are like magicians; and just as we should understand the trick before answering whether or not the magician has really made an elephant vanish into nothing, so too we should refrain from supposing that simulated intelligent behavior is truly intelligent until we know the trick. We may of course be deceived about real frogs—they may be little more than fleshy robot-frogs. I cannot seriously entertain the possibility that I might be deceived about myself: I think, therefore (at a minimum) I *think*. I may generate the solutions to my problems by way of unknown mechanical processes, but surely I grasp the problems and I grasp the solutions as solutions to those problems. My belief in the reality of intelligent behavior is a belief in the reality of problems.

I do not pretend that by mimicking Descartes's *cogito,* "I think, therefore I am," [35] I have (re)produced a first principle for philosophizing about the reality of intellect. But I do wish to maintain that genuine intelligence affords an alternative to instinct, a hypothesis contrary to that of innate, pseudointelligent behavior. Unlike Descartes,[36*] moreover, I am fully prepared to suggest that other animals lower down on the evolutionary ladder (or merely lower down on the food chain) may also have genuine intelligence, expressed in such lowly pursuits as fly catching.

2.14 Dualism, Mechanism, and the Hypothesis of Genuine Intelligence

The hypothesis of genuine intelligence seems more to demand an explanation than to supply one—something AI theories might be said, in principle, to do. But although this hypothesis might even seem to remove the ground of physical support from the scientific image of (genuine) intelligence, it is too early to conclude that the hypothesis has perpetuated an untenable dualism between physical reality and an incorporeal thinking substance, the intellect. The sort of argument Descartes relied upon to draw his mind-body distinction would be unavailable with respect to the intellectual faculty here hypothesized of (among others) the frog; for without fully conceptualizing the items in its experience, the frog grasps and solves problems in ways that demand real physical movements and coordinations of those movements with physical objects. Intelligent behavior of this sort demands the presence of the corporeal world in ways that Descartes's mere doubts about the world do not.[37*]

So it might be argued that a pernicious dualism is denied any theory that incorporates the hypothesis of genuine intelligence, for such intelligence is within the world, not divorced from it. But there is still the danger of a kind of dualism by default, a segment of the manifest image that can't be reconciled with the scientific image of animals. For as scientific inquiry presses in the direction of a biological theory of intellectual functioning, it is likely to discover numerous mechanisms that account for some features of that functioning; yet it is far less likely to detect sources of control, problem grasping, and so on, which, by hypothesis, transcend those mechanisms. Of course if science discovers enough mechanisms (including mechanisms that operate *as if* problems were grasped, etc.) to account fully for what we presently regard as the intelligent functioning of the organism, then real intelligence is an illusion, and our personal sense of grasping a problem is a mere epiphenomenon, functionally insignificant.

Is there any possibility of a nonmechanistic but still scientific theory of intellectual functioning? If a mechanistic theory is just a deterministic theory, then the obvious alternative would seem to be an indeterministic theory; but all that indeterminism seems to do is introduce an element of chance into the story of intellectual functioning, and chance is hardly adequate to explain intelligent problem solving.

This is a familiar line of argument in connection with the issue of free will, and there the point is made that chance is not adequate to explain our intuitive notion of free will: Making a truly free choice is obviously different from, say, flipping a coin. Asked why this is so obvious, we might not be able to give a compelling answer, since our sense of a truly free choice seems rather dim. (A typical answer is that free will demands determinism in order to sustain causally efficacious choices, but this grabs the problem from the wrong end; this answer concerns itself with events taking place after the choice, not those leading up to the choice.)

In the case of truly intelligent functioning, our sense of the process is not quite so dim. We have some awareness of events (problem grasping, etc.) leading up to intelligent solutions to problems. Given that awareness, we are almost forced to admit that merely introducing an element of chance into the problem-solving process, although sometimes a useful alternative to a plodding approach to the task of finding a solution, would do little to account for grasping a problem, recognizing its solution, and so on. The AI approach to a theory of intelligent functioning could easily introduce some chanciness, say, by making randomizing devices an integral part of its heuristic programs for problem solving; but, notwithstanding also programming the computer to print out "Oh, look: a problem!" before it blindly carried out the heuristic programs, those programs could hardly be said, by virtue of their element of chance, to have come any closer to the requisite grasping of problems and solutions.[38*]

Chapter 3

Two Psychophilosophical Theories of Intelligence

This chapter examines two roughly similar theories of intelligence that differ sharply about whether intelligence is an explanatory alternative to instinct. The two theories, Köhler's and Dewey's, are philosophically proficient but, judged by the methodological standards commonplace in contemporary psychological research, not scientifically well tested (though Köhler's Gestalt theory is based, however loosely, on a wealth of suggestive observations and experiments). Nonetheless, like Tinbergen's theory of instinct (or, for that matter, neo-Darwinian theory [1*]), the theories in question may afford valuable ways of looking at a host of phenomena, ways of thinking about empirical findings.

3.1 Reintroducing Gestalt Theory:
Koffka on Instinct as a Prelude to Intelligence

A useful introduction to Köhler's decidedly instinct-oriented theory of intelligence might be provided by Koffka's Gestalt-theoretic remarks on instinct, some of which have already been noted (see sections 2.3 and 2.4). Arguing against the view that instinct may be understood in essentially mechanistic terms, as chained reflexes, Koffka cites some findings about one spider's bee-fleeing instinct: "The effective elements of stimulation must . . . differ in accordance with the position the bee occupies with respect to the spider. Yet movements of flight are released even when the bee occupies the most unusual positions. Here we have an endless number of possibilities of stimulation by the same object; consequently, if the instinct-apparatus is conceived as a system of predetermined paths, these pathways must be almost infinite in number." [2]

Despite the rejoinder that there may be a small number of commonalities (e.g., buzzing, rapid flapping, yellow and black, etc.) among the seemingly endless number of specific stimulus occasions, this objection has some undeniable appeal. For even if (say) any yellow-and-black pattern were to set off the spider's fleeing reaction, the variety of possible yellow-and-black patterns would be as difficult to account for (indeed, difficult for comparable reasons) as accounting for the variety of bee positions.

Koffka's alternative to the chained-reflex theory of instinct begins by suggesting that instinctive activities "possess the same forward direction that is characteristic of voluntary action."[3] But Koffka worries about the puzzling way that goals "of which nothing is known"[4*] may direct an animal's course of activity. His solution is characteristically Gestaltist: The psychical field is a dynamic one striving for a kind of equilibrium. The present need situation "is not a *state*, but a *transition;* not a *being*, but a *becoming*."[5] Like "the tragic end which hangs over the audience"[6] from the first scene onward, like the yet-to-be-played ending of a melody, the outcome (or goal) of instinctive action is foretold in the configuration of the psychical field. Organic need causes the instinctive activity; and "not only the need, but also the way in which the need is satisfied, appears to be inherited."[7] The need prompts activity, which proceeds in the direction of the satisfaction of need (or the relief of dynamic tension), toward a kind of closure. There is an inner, affective as well as outer, overt-behavioral aspect of instinctive activity: "Furthermore, these emotions, this 'inner behavior,' will fit the external behavior of the instinctive act perfectly, just as our general conception of behavior requires that they should."[8] A perfect fit, I might add, need not be a functionally isomorphic one: The felt need may, on Koffka's view, precede the overt behavior; and "the general direction of one's total behavior is all the time determined by more potent forces which are not projected into consciousness."[9] The "perfect fit" would seem to be a matter of exact, functionally appropriate correspondence whenever an inner aspect of instinctive activity corresponds at all.[10*]

Notwithstanding differences between instinctive and intelligent behaviors, Koffka views them as fundamentally similar. He hopes "to give an ultimate explanation [for both] . . . in terms of the universal law of *Gestalt*."[11] Habit, too, is said to be an expression of this same law or principle, "the principle of configuration."[12] Here as elsewhere Koffka's views conform closely to Köhler's, so let us turn, as if in consideration of further details of the same account, to Köhler's theory of intelligence.

3.2 The Problem of Intelligence: To Grasp Ideas as Solutions

According to Köhler, "*Understanding* or *grasping* is the first important kind of intelligent behavior."[13] This grasping is said to range from passively taking in to inventing or discovering. Köhler argues that the possession of ideas cannot account for such mental acts, at least not entirely, since ideas per se would not necessarily change the situation as all the varieties of grasping do.[14*] In support of his argument, he asks: "Does the profusion of ideas in 'flights of ideas' illuminate processes of understanding or invention? It hardly seems so. And who does not know one or another of those awful people who bury problems and situations under an avalanche of inappropriate concepts rather than bring them to a solution?"[15]

One apparent corollary of this argument is that the presumption of innate ideas cannot, by itself, account for intelligence. Köhler himself spells out a related corollary about a theory somewhat analogous to Darwinian theory, a theory that applies ontogenetically rather than phylogenetically to the adaptive behavior of certain animals:

> In a new situation, it is said, for which there is no ready-made behavior, [some] creature produces all kinds of chance responses. Some sequences of such responses . . . would by chance be such as to constitute objectively suitable behavior precisely for this particular situation. A very practical arrangement in the nervous system provides that those sequences of responses which lead to success become physiologically fixed and acquire a tendency to be repeated in similar external situations, and adaptation thus comes about.[16]

Köhler next describes a variant of this theory, one that tries to explain intelligent behavior on the basis of further mechanisms of chance variation and selection at the level of ideas. Instead of producing chance variations in overt behavior, the intelligent creature is alleged to produce them internally, as sequences of ideas: "Most of these randomly produced sequences will, of course, not be adaptive; but a few of them will be, will be translated into reality, and then will be stamped in for similar reasons."[17] Köhler then argues, in effect, that far from obviating the need for an animal to grasp its situation, this theory, if it is to live up to the claims made for it, must posit a principle of selection that involves such grasping. The earlier version of the theory had the disadvantage of merely allowing for the selection of solutions actually implemented in behavior. To preserve its advantage over the earlier theory, the idea-version of the theory must also allow for selection based upon the organism's intellectual grasp of the situation: "The individual compares, so to speak, the chains of ideas arising randomly in him with his perception of the situation, and . . . in this process he is able to grasp which sequences of ideas, if converted to action, would fit the situation objectively, and which would not, which, in other words, might be considered for actual execution."[18]

Insofar as one grants that the main issue is not "the actual production of ideas as material, but whatever it is that is called insight," Köhler suggests that one has "conceded that the problem of intelligence concerns a type of function, a grasping of relatedness within the material."[19]

An attempted "short-cut" solution to the problem of intelligence might suggest that the requisite "relatedness" is *given*, that animals with higher intelligence "have not only what is called the 'contents' of consciousness in the narrower sense, but also in addition all relations among all such contents."[20] But Köhler argues against this suggestion as contrary to actual and to possible experiences: "Since I know very well what it means to be conscious of a really existing relation when it occurs, I must emphatically deny that I would have some hundreds of such relations simultaneously, or even that I could do so."[21] Now some critics might claim to be far more proficient than Köhler at mental juggling. But once it is pointed out to them that his estimate of hundreds is too conservative, that the number of relations among contents of the perceptual field is potentially infinite (that there will be, e.g., relations among relations; relations among relations among relations; and so on),[22*] these critics should concede Köhler's point.

Köhler then argues that even if the suggestion were true, it would not help us to understand the function of intelligence: Merely having the relations in mind would not be enough to account for understanding and invention. If one had them in mind, one also would have to decide which of them suited one's situation: "And so this theory, which would offer us too many relations, only leads back once more to a principle of selection."[23]

3.3 The Supposed Continuity of Instinct and Intelligence

Innate ideas are also not the key to the problem of intelligence, where intelligence is a matter of going beyond any evolved patterns of behavior that fail to fit novel environments. This conclusion, like others expressly drawn by Köhler, is based on the consideration that a ready-made supply of ideas would need to be sorted through, that appropriate ideas would have to be selected from among the available supply. Such a selection process would require a grasp of the ideas and the situations to which they apply.

But what of any evolved patterns of behavior that do fit certain environments? It might be argued that such patterns would have been preselected during phylogeny and, so, would not stand in need of any further processes of selection during ontogeny. The "potent forces . . . not projected into consciousness"[24] that determine the direction of instinctive activity are, according the Koffka, innately selected principles of configuration. Yet assuming that the same forces serve to direct intelligent activity as well, it could be suggested that the difference between instinctive and intelligent configurations is that the former are fully selected (for implementation) in Darwinian fashion, whereas the latter are further selected, from among a repertoire, on the basis of insight.

Despite its plausibility, though, this suggestion conflicts with Koffka's considered opinion that there is no sharp line between instinct and intellect: "We can not say that intelligence begins where instinct leaves off; for this would over-emphasize the inflexibility of instinct. Studies of the instinctive behavior of insects show that there is little or nothing of clock-work precision about it, and our criterion of intelligence has its application here just as it does to the behavior of man."[25] The criterion mentioned is flexibility; and, so, if flexibility calls for principles of selection during ontogenesis, then both instinctive and intelligent behaviors call for such principles.

In claiming a fundamental continuity between instinctive and intelligent behaviors, the Gestalt theorists might be thought to have sided against instinct-oriented explanations of behavior—especially since that claim is predicated upon the flexibility of both sorts of behaviors. But despite the problems that flexibility (or elasticity) poses for some efforts to provide a criterion for innate behavior (see section 1.8), the Gestaltists might seem to have found a way to circumvent any such problem; for they purport to explain that flexibility in terms of dynamic laws. By positing such laws, Köhler and Koffka commit themselves to a sort of biological determinism that is favorable, if not tantamount, to a theory of instinct.

3.4 Free Dynamics, Particular Constraints

Köhler contends that dynamic laws play a fundamental role not only in psychological systems, but in biological systems generally. Distinguishing between "the free dynamics . . . which brings about the change and determines [the] direction" of a system and the "particular constraints by which a special form of [a system] is prescribed," [26] Köhler observes that the mechanistic constraints upon biological systems are in a state of flux, that even if some mechanistic laws are immutable, their material supports vary:

> No part of the anatomy of an organism is a permanent object. Rather, any such part must be regarded as a steady state, only the shape of which persists, while its material is all the time being removed and replaced by [way of] metabolic events. Surely, such steady states are fairly resistant and can, therefore, serve as constraining devices for more temporary functions. But from the point of view of our general topic, it is most important that all organs, large or small, are processes rather than permanent objects.[27*]

Köhler hopes to explain even the self-maintenance of mature organisms in dynamic terms, and he compares this signally important trait of living systems to the dynamics of a candle's flame: "When a flame, say that of a candle, has developed to maximal size, it maintains itself at this level so far as it can." [28] Mechanistic laws do not, according to Köhler, tell the whole story of living systems—though this does not imply that such laws do not tell an indispensable part of the story. His characteristically holistic tendencies find one expression in his "general rule that a system in which local events depend upon translocal dynamic interrelations can only be understood and handled scientifically as a whole." [29] On the reasonable premise that intelligent behaviors and their psychical accompaniments are "local events" of this sort, Köhler concludes, speculatively, that they "are based upon processes in our central nervous system set up by outer stimulation of the sense organs, or by inner stimulation, and that these underlying processes represent only a fraction or province of a larger physiological context in our organism." [30*] Such considerations lead him to view psychology as a branch of physiological biology and, in turn, to view that biology as subsumable, ultimately, under physics.

Restricting attention to the first two levels of description (of local events and physiological systems), Köhler assumes a dynamic, "mutual influence between local events." [31] That influence, which is not wholly evident at the level of those events, can begin to be appreciated only at the level of the underlying physiological system, understood by Köhler as a dynamic rather than entirely mechanical system. This system—which encompasses (functional counterparts of) sensory input, intelligent behaviors, and their psychical accompaniments, as well as other "local events in the working of the nervous system"—is subject to dynamic tendencies toward self-(re)arrangement.

3.5 A Dynamic Model of Attention

An approximation to Köhler's view of intelligence is available in his account of attention, which is a sort of grasping and, so, by his lights, an exercise of intelligence. Köhler stresses the need to treat attention within its proper context: "There is no such experience as attention taken separately, apart from any objective; rather a person feels directed towards something, in the manner which is called attention." [32*] Köhler hopes to unite this "functional" notion of attention (and other "attitudes" such as striving and fearing) with his theory of the "field structure" of spatial experience:

> Attitudes issuing from the self and directed toward definite parts of the environment will thus be regarded as based upon directed states of the field. These directed states must exist between the processes underlying the self and the processes corresponding to those definite things toward which the self is directed. Whether we experience such directed attitudes or not and whatever their direction and their qualitative nuances are, depend upon the actual properties of the self and the actual properties of our environment. [33]

On this view, the dynamic equilibrium of our system of experience is disrupted by something within ourself or our environment; and the dynamic laws of that system, which tend to reestablish a dynamic equilibrium, give rise to a directed state of the field itself—to a directed attention, striving, or other experienced attitude. Koffka mentions a physical analogy used by Köhler to clarify his account:

> A soap-film is produced upon a wire-frame, and upon it a little noose of thread is cast in whatever form it may take. If one proceeds carefully, the thread will be supported upon the surface of the film, "but if one pricks

the film *inside* the noose with a needle, the surface will break apart and the thread will be pulled out by the surface-tension of the outer portion of the film, which seeks to give the area outside the thread the least possible surface, and the area circumscribed by the thread the greatest possible surface. As a result, the thread immediately assumes the form of a circle." In this example we can conceive of circularity as the "end-situation," puncturing the soap-film as the stimulus releasing the movement, and the movement itself as the "transitional situation." [34]

The soap film might be compared to our organized field of experience, consciousness of the world. Some items might enter that field without disrupting its dynamic tension, but other self-initiated or environment-initiated factors might serve as pricks to consciousness, causing the field to reorganize itself around, direct itself toward, the new items. Dynamic laws governing the transition move the field toward some (potentially predictable) closure—an "end-situation" that is a newly reorganized field of objects as experienced.

The extended analogy, with its noose of thread, might seem to suggest, wrongly, that attention inevitably reorganizes its objects—or, at any rate, seems to be best suited to cases of not very stably organized objects of attention, objects such as Figure 3.

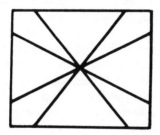

Figure 3.

Köhler says that "after its center has been fixated for some time, [this figure] usually changes suddenly from one organization" to another.[35] But by varying certain features of the soap-film example, by proposing that a fairly rigid wire be placed on the soap film, one could use the analogy to convey the idea that some objects of attention might be stable enough to resist any reorganizing tendencies of the field of experience. Thus, one putative bias of the analogy is corrected by means of an implicit appeal to Köhler's own distinction between *free dynamics* and *particular constraints*.

Köhler's overall account of attention is very attractive and has notable internal resources for meeting objections. If we nevertheless resist his soap-film image, which is after all only an analogy, perhaps we are merely registering a reluctance to put ourselves in the place of inanimate objects, to see ourselves as scientists tend to see us, objectively. Should neurological counterparts to that analogy ever be discovered, understanding ourselves in Köhler's dynamic, Gestalt-theoretic terms might become easier than it is now.

3.6 Goal Seeking as Stress Directed

Köhler attempts to extend his account of attention to some activities often deemed instinctive but also construable as intelligent, to food-seeking behaviors. The spur to action, like the prick to consciousness, is a disruption of dynamic equilibrium. The source of disequilibrium emphasized by Köhler is the internal condition of need for nourishment, a condition experienced as hunger:

> No outer physical force drives the physical organism toward food when food is needed. But in the brain field there is a directed stress when the self-process is altered in the way we experience as hunger and when, among the environmental processes, there is one complex which we experience as food. The stress, of course, exists in that region of the field which is functionally between the self-process and the food process. Its tendency is to decrease the functional distance between the two.[36]

In cases where food is within the field of experience, hunger spurs activity directed (or "steered") dynamically toward the food: "Among all operations of the muscles possible at the time, certain constellations result in the locomotion of the physical organism toward the physical food outside.[37] Hunger might not actually initiate the activity that becomes food seeking, since the organism might already be active when hunger strikes (in which case the hunger might simply serve to redirect the activity toward food in the environment).

A precondition of conscious steering toward food might be the awareness of something as food; and the stress of felt hunger might reorganize the field of perceptual experience accordingly, might lead to a special kind of attentive regard toward perceived objects, to an awareness of them as food. But what if, perchance, there is no suitably nourishing material in the field of experience? In that case, free dynamics might override the particular constraints of perceptual objects. Consider, for example, Charlie Chaplin's predicament of near starvation in the film *The Gold Rush*: At one point, his equally hungry cabin mate sees him as a man-sized chicken. At another point, both men perceive a shoe as a suitable ingredient for soup.

When no even slightly plausible foodstuffs are at hand, then changing the field of experience seems a good recourse for the hungry animal. Locomotion might

afford a means to a more satisfying field of experience, but what directs the locomotion? Initially, perhaps nothing does. Whether prompted by the spur of hunger
or simply an inertial state of the organism, the locomotion might be random until
a new field of experience coalesces, attentively, around a new item of experience.
Then, when the animal becomes conscious of the item as food, further activity
might be steered, as is customary in such circumstances, toward that item.

3.7 Beyond the Present, toward the Absent

Köhler and Koffka were not fully satisfied with their own position, which they
put forth pending closer physical analogies to the sorts of dynamic principles they
presumed to be necessary for the explanation of psychological systems. It was
not just that no neurological counterparts to the psychological dynamics had been
discovered, but also that no sufficiently temporal analogy had been proposed. The
soap-film analogy was considered too static, the events too independent of time,
to account for instinctive activity over time. As Koffka put it: "The unitariness
of instinctive action very obviously suggests that with 'closure' the whole temporal course of the activity is involved, and, so far, we have not referred to any
dependency upon time." [38]

Of course the soap-film analogy is not exactly independent of time, since the
dynamic process of re-equilibrating around the noose of thread is patently temporal. But perhaps the Gestalt theorists have wrongly characterized their own
desideratum as one of paying due heed to the temporal component of the behaviors they hope to explain. It is not so much that instinctive and/or intelligent
activity goes beyond *the present time,* but that such activity sometimes goes beyond *what is present within* the present field of experience, what is *presented
to* the active organism's consciousness. Harkening back to the metaphor of the
unfinished tune whose configuration gives rise to an almost palpable sense of closure, we might say that the Gestalt theorists wish to view the whole of instinctive
activity as governed by the inner melody whose end is unknown and absent. It is
the dynamic tendencies (of instinctive and/or intelligent behaviors) toward absent
goals for which Kohler and Koffka can find no clear physical parallel.

The need for such a parallel might be questioned on the grounds that no such
tendencies have to be posited to explain food-seeking behaviors (among others).
Thus, as I suggested above, a hungry animal's locomotion in the absence of foodstuffs might not be directed but random, yet still achieve a potentially satisfying
outcome, a field of experience containing edible things. But this suggestion is not
likely to appeal to the Gestaltists, for it amounts to a denial of their key contention that innately scored melodies (or principles of configuration) fully guide our
instinctively intelligent and/or intelligently instinctive behaviors, note by note,
toward their harmonious completions. All that I presently wish to suggest is that

this particular contention could be abandoned without giving up on the whole Gestalt theory.

3.8 Dewey's Naturalistic Standpoint

Although one finds a remarkable degree of accord between Dewey's theory and the Gestalt theory, there are some fundamental disagreements, too. The most important difference, for present purposes, is that Dewey's theory is far less nativistic than Köhler's. Both agree that our fields of experience are organized, but while Köhler attributes that organization to innate principles of configuration, Dewey chalks it up to habits, which would seem nonnativistic to their core.

Dewey's principle of habit owes much to William James's law of habit. Jamesian habits, like Gestalt principles of configuration, are housed in the nervous system. The basis of habit is said to be the plasticity of and "the growth of structural modification in living matter,"[39] especially nervous tissue: "Our nervous systems have . . . *grown* to the way in which they have been exercised, just as a sheet of paper or a coat, once creased or folded, tends forever afterward into the same identical folds."[40]

Unlike James, Dewey makes no effort to reduce psychological laws of habit to physiological ones. Unlike Köhler, Dewey does not suppose that the nervous system contains functional correlates of "actual properties of the environmental processes"[41] that affect the organism. Dewey's alternative, naturalistic standpoint is this: "Living as an empirical affair is not something which goes on below the skin-surface of an organism: it is always an inclusive affair involving connection, interaction of what is within the organic body and what lies outside in space and time, and with higher organisms far outside."[42]

Dewey sees no merit in housing functional replicas of the environment either in the mind or the nervous system. Experience, he says, is not in the mind or the nervous system, but in nature:

> The soul was conceived as inhabiting the body in an external way. Now the nervous system is conceived as a substitute, mysteriously within the body. [There is an] intimate, delicate and subtle interdependence of all organic structures and processes with one another. . . . To see the organism *in* nature, the nervous system in the organism, the brain in the nervous system, the cortex in the brain is the answer to the problems which haunt philosophy. And when thus seen they will be seen to be *in*, not as marbles are in a box but as events are in history, in a moving, growing never finished, process.[43]

This naturalistic though nonreductionistic view of ourselves and our experience informs Dewey's principle of habit (or continuity of experience): "The basic characteristic of habit is that every experience enacted and undergone modifies the one who acts and undergoes, while this modification affects, whether we wish it or not, the quality of subsequent experiences. . . . The principle of continuity of experience means that every experience both takes up something from those which have gone before and modifies in some way the quality of those which come after." [44]

Dewey does not so much deny that experience is internal to the organism as he asserts that it is also external, that it involves interaction between the organism and its environment. When he speaks of the continuity of experience, he makes reference to the stability of changes not only in the organism but also in its environment. Experience itself is essentially active, and, insofar as it involves doing things in the world, experience changes the world. The changes wrought then continue to affect subsequent experience, so habit, the persistence of experience, is as much in the environment as in the organism. [45*]

3.9 Habits

No other concept is more signally important to Dewey's thinking about human beings than *habit*. Habit has so strong a hold on our behavior, so much command over us because "we are the habit." [46] In fact, instead of viewing habits as things that hold sway over us, against our will, he proposes the neat equation that habits "*are* will." [47] They are the very means by which we act, not merely the tools or materials used for certain purposes: "They are active means, means that project themselves, energetic and dominating ways of acting." [48] Habits may not be identical to our wishes or ideas; but, according to Dewey, "a wish gets definite form only in connection with an idea, and an idea gets shape and consistency only when it has a habit back of it." [49] Even if ideas or purposes have been hit upon by chance, habits are the requisite mechanisms whereby those ideas are "carried into execution." [50*] Our perceptions are organized by habits; our sensations can be discriminated only on the basis of them. Habits give rise to meanings; habits are the "intermediates, middle terms" that put us in touch, whether in thought or action, with our ends. [51] Our very character would not exist save "for the continued operation of all habits in every act": "Character is the interpenetration of habits." [52] The list could go on, but I will end it with Dewey's observation that "immediate, seemingly instinctive, feeling of the direction and end of various lines of behavior is in reality the feeling of habits working below direct consciousness." [53] It is an exaggeration, though just barely, to say that any psychological phenomena not directly accounted for by Dewey in terms of our habits are, instead, indirectly accounted for in terms of the breakdown of those habits.

Habits are ever operative:

The habit of walking is expressed in what a man sees when he keeps still, even in dreams. The recognition of distances and directions of things from

his place at rest is the obvious proof of this statement. The habit of locomotion is latent in the sense that it is covered up, counteracted by a habit of
seeing which is definitely at the fore. But counteraction is not suppression.
Locomotion is a potential energy, not in any metaphysical sense, but in the
physical sense in which potential energy as well as kinetic has to be taken
into account of in any scientific description. Everything that a man who has
the habit of locomotion does and thinks he does and thinks differently on
that account.[54]

The point is, as the principle of continuity says: Experiences live on, continuing
to influence subsequent experiences, which involve doing things. In this sense,
habits are always operative, though not necessarily in the same form of activity
commonly associated with those habits. Thus, the habit of walking should not
be confused with the activity of walking, which activity is but one expression of
that habit. Another expression of that habit is the recognition of distance, which
recognition is said to depend on prior experiences of walking (and other forms
of locomotion). It would, of course, be a grave misunderstanding of Dewey to
maintain that his suggestion about the actuality of all habits a person has amounts
to the supposition (say) that if a motionless person has the habit of locomotion,
that person must be locomoting in thought. Indeed, the actuality of habits, known
mainly by its effects, may sometimes have very little to do with anything—either
mental or physical—inside ourselves: Sometimes our habits might be said to
subsist principally in the physical or social environment. And saying this is no
more or less an exaggeration than saying, instead, that our habits can endure,
covertly, in the mind. Habit is a property of experience, is its continuity; and
normal experience is an interaction or transaction between internal conditions
("personal needs, desires, purposes, and capacities"[55]) and objective conditions
(the physical and social environment): "Taken together, or in their interaction,
[these conditions] form what we call a *situation*."[56] The locus of habit, its always
current dwelling place, is this interaction, the situation incorporating the ongoing
participation of both sorts of conditions.

3.10 Dewey's Seamless Web: Experience without Representation

Dewey's "situations" correspond roughly to what Köhler would call "fields of
experience," but there is this signal difference: Köhler locates the field of normal
experience, that is, experience of the world, within the organism; Dewey does
not. Following an inveterate tendency of many philosophers, Köhler posits *representations* of shared reality. Such representations, whether housed in the mind
or (more fashionably) in the brain, are supposed to be constituents of our fields
of experience of the world.

One might hope to characterize the difference between Dewey and Köhler by saying that for the latter but not the former it would be possible for us to have the same experience we now have of the world were we to be mere brains in a vat of chemicals and to be suitably stimulated, chemically or electrically, by a mad scientist. But this characterization is not quite right, as the following metaphorical comparison indicates: Both theorists might be said to view principles of organization of experience as a complicated web extending from the environment to the innermost recesses of the organism. Köhler's theory, unlike Dewey's, posits a seam (in the web) within the organism. This seam is the field of normal experience. The suggestion about the brain in the vat was unfair to Köhler, since on his view as well as Dewey's, disconnecting the web from the environment quickly leads to the collapse of the web and, hence, the collapse of experience of the world.

The figurative counterpart of the brain-in-vat hypothesis would be the idea that some other support for the web might be found; but the parallel is not exact, since the hypothesis does not construe matters in appropriately dynamic terms. There is on the brain-in-vat hypothesis the possibility of intercepting all natural messages from the environment and sending phony messages instead. Viewed in causal terms, this amounts to interrupting the sequence of physiological events normally extending from the peripheral sensory receptors to the central nervous system "headquarters" in the brain. The end of that sequence could, in principle, be exactly reproduced by initiating (chemically or electrically) identical events at the point of interruption of the normal sequence. But consider the lack of parallel in the case of the (dynamic) web: Suppose we sever its connections with the normal environment and reconnect it at some point in closer proximity to the seam. We have then changed the dynamics of the system, and the seam will not seem the same: The web vibrates differently when its individual strands have been shortened.

Can we duplicate the dynamics of the normal system while severing its usual connection to the environment? One way of doing so would be to reattach the strands, without shortening them, to a fake environment that resonates identically to the real environment. This trick, like replacing one Stradivarius violin with another that shared all the Strad's occurrent properties, scarcely qualifies as a trick. A trickier trick would be to shorten the strands yet compensate, somehow, for the shortening. This is the closest analogy, on Köhler's view of the dynamic psychological system, to the brain-in-vat hypothesis. In the dynamic neighborhood of the seam, such a trick would have to—and conceivably could—produce effects indistinguishable from those normally wrought by the normal system. It is arguable that on Dewey's view, in the absence of any seam, the proposed change in the overall dynamic system would amount to a direct change of experience, that no such trick could possibly work.

Do we need the hypothesis of a seam? Well, if that means, do we need a field of experience? then of course we do. But if the question asks instead whether we need a seam to account, philosophically or scientifically, for the field, then Dewey may be understood to have answered no. My imagery is a bit unfair, since obvious seams might be thought to be a sign of poor workmanship, and Köhler's theory is not poorly sewn together. What he proposes is more like a duplication of the larger dynamic system of the organism and the environment, within the brain of the organism. Moreover, the smaller system is conceived to be not merely miniaturized but radically transformed, preserving (in the case of perception of things in space) only functional relationships among counterparts of items in real space that also bear geometric relationships among themselves. Still, the question remains: Do we need the hypothesis of this special representational system? If we cannot make sense of a field of experience in the larger system, then we will not be any better off trying to do so in the smaller system. If we can make sense of the field, as Dewey tries to do, in the larger system alone, then the smaller representational system is an unnecessary multiplication, duplication, of reality.

3.11 Figuratively Stirred to Consciousness

The web metaphor conveniently reifies the dynamic principles of configuration—which, on Dewey's theory, are habits—as individual strands of the web. The idea of a web conveys a vivid sense of an overall dynamic equilibrium of a system of habits. As our environment changes, our habits may be flexible enough to remain connected to objective conditions. If not, a rupture of the strands occurs and the whole system, a living system, becomes aware of the breakdown. Insofar as the web is an integrated network, a dynamic system of strands, every strand influences and is influenced by what happens locally, at the points of rupture, where the field of experience may be said to have its principal foci. Suppose we view the field of experience as the set of vibrations occurring in response to the breaking of a strand of webbing. Then it may be said that each interconnected strand contributes to the resultant set, even when the particular strand, not having snapped, is not the original source of the vibrations. Moreover, the vibrations of each of these other strands will resonantly affect the character of the vibrations of the strand that broke. Recognizing, as it were, the need for greater flexibility, the broken strand becomes more elastic, stretching and flailing about as if in the hope of a rapprochement with the environment.

Since the idea of a living web that reconstructs itself seems hard to reconcile with nature's own examples of webs, let us reify agency in the natural form of a spider. Picture the spider sitting impassively at the center of the dynamic network of strands comprising the web. Each strand is the material support for habitual action. Virtual tropism may ordinarily suffice to impel the spider toward the point of origin of vibrations—say, a gnat caught on one of the strands. Let us assume that the spider's behavior as it locates and devours the gnat is (largely) unconscious. Occasionally, though, more major perturbations of the web occur: The massive vibrations signal the rending of strands, the breakdown of habits. Fully conscious, the spider again moves, perhaps more rapidly, toward the source of vibrations and finds strands disconnected from a branch permanently bent away from the web, as a result of some uncaring mammal's footstep. Out of his own substance, the spider spins new strands, reconnecting old ones to the newly located branch. Guided by old habits (the still intact portions of the web), the spider tries, as far as possible, to preserve the original design. All reconstruction in novel circumstances is modified by environmental constraints, however, so the new web, adapted to an altered environment, is different.

At the level of normal experience, of active involvement with the environment, the reverberations of habit (of all habits) are felt. We are not aware of the habits per se, yet our normal experience of the environment is affected (and effected) by their vibrations. The experienced familiarity of a face; the cognizance of the meaning of a written word without awareness of the shape of its letters, the color of the ink, and so on are instances of what Dewey calls the influence of habit on experience. The fact that we are not directly aware of these workings of habit does little to call their existence into question. Dewey's suggestion that habits are actively there, all along, may be viewed as an important reminder that the whole story about the manifest image of human beings in the world is not tellable exclusively in the terms of that image, that we sometimes have to go beyond experience in order to understand it. We transcend ordinary experience as we enter a realm of theoretical posits that may or may not be integrated into a fully articulated, directly testable abstract structure, the scientific theory. Now it might seem an act of intellectual hubris for philosophers to examine—let alone to hypothesize, as Dewey does—such a realm, but they might still be able to distinguish posited hawks from posited handsaws and to assess the plausibility of the stories told about each.

3.12 Impulse as against Instinct

Dewey, like Köhler, explains the activities of living things in dynamic terms, as efforts to reestablish a stable equilibrium after a state of *dis*equilibrium called "need." Dewey's point of departure from Köhler is the insistence that although organization is an undeniable fact of organic activity, there are no innately organized or organizing tendencies of this activity. Dewey does however posit some unorganized native tendencies, which he calls "impulses." Although he uses "impulse" and "instinct" as "practical equivalents," Dewey observes that the two terms are importantly different in meaning: "The word instinct taken alone is still too laden with the older notion that an instinct is always definitely organized and adapted—which for the most part is just what it is not in human beings. The word impulse suggests something primitive, yet loose, undirected, initial."[57]

Among the facts that incline us to speak of a hunger instinct are, according to Dewey, that: "The paths of motor outlet or discharge are comparatively few and are fairly well defined. Specific bodily organs are conspicuously involved. Hence there is suggested the notion of a correspondingly separate psychical force or impulse."[58] But Dewey criticizes this suggestion on the basis of other facts, first, about the organism and, second, about its environment. First:

> No activity (even one that is limited by routine habit) is confined to the channel which is most flagrantly involved in its execution. The whole organism is concerned in every act to some extent and in some fashion, internal organs as well as muscular, those of secretion, etc. Since the total state of the organism is never twice alike, in so far the phenomena of hunger . . . are never twice the same in fact. . . :
>
> In the second place, the environment in which the act takes place is never twice alike. Even when the overt organic discharge is substantially the same, the acts impinge upon a different environment and thus have different consequences. It is impossible to regard these differences of objective result as indifferent to the quality of the acts.[59]

The argument seems to be that the actual fit between organisms and their physical surroundings is far too variable to warrant our attributing it to a preestablished harmony of instinctive activity. Still, Dewey is willing to concede that, in the case of some animals other than humans, some original activities do qualify as instinctive. Consider, for example the posthatching behavior of chicks that are zoologically classified as *precocial* (i.e., down covered and, more importantly, actively able to fend for themselves). Although such chicks are, as organisms, wholly involved in particular acts yet not wholly the same over time, and although their environment is constantly changing, the chicks' posthatching pecking is equal to the task of satisfying their nutritive demands, their hunger; and Dewey is prepared to classify their pecking as instinctive.

Somewhat surprisingly, given Dewey's laudable efforts to stress humanity's place in nature and his usual tendency to find continuities beneath apparent dichotomies, Dewey sees a rather sharp contrast between humans and brutes. Although he affirms that "instincts in the animals are less infallible and definite than is supposed," he contends that "the human being differs from the lower *animals* in precisely the fact that his native activities lack the complex ready-made organization of the animal's original abilities."[60]

In urging humanity's lack of initial organization, Dewey appeals to our great helplessness at birth. But compare human beings to those avian species that zoologists classify as *altricial* (that is, helpless at hatching and, so, in need of parental care and feeding). Instead of pecking, these birds exhibit gaping behavior, that is, mouth opening in the general direction of obliging parents. This behavior seems to be a rather complex coordination of the hatchling's behavior with that of its parent. The phylogenetic background of all this is far from obvious. Suppose the behavior was selected for in a distant past environment in which red sap from (now extinct) trees would drip into the mouths of hungry hatchlings. Later, as the trees grew scarce, some parent birds feeding nearby happened, quite by accident, to drop a morsel of food down the gullet of some delighted hatchlings. A few hatchlings of this sort might have lived to carry on the tradition—not because of a dual process of selection, but because of an already selected tendency of animals to repeat the pattern of rearing to which they have (perhaps unwittingly) been exposed by their parents. Thus it seems that precocial behavior could become altricial even without any further evolutionary modifications of the organism. Gaping is all too apparently more helpless than pecking, but helplessness has more to do here with the character of the environment than with that of the behavior. Consider another example: Some altricial hatchlings peck at a red spot on their parent's bill, producing a response of regurgitation into the hatchlings' gaping mouths. Once upon a time, to spin another fanciful yarn, this pecking may have loosened red berries from trees of a species now extinct, and the subsequent gaping may have served to catch the berries when they finally fell. As the trees grew scarce, a hapless parent might have been surprised to receive a nauseating peck on its bill. Not very bright, the parent repeatedly flew off, ate more food, then returned to the nest. Later, when the hatchling grew up and had hatchlings of its own, it iterated, as animals often do, the rearing practices of its parent. Another tradition was born.

By the same token, a precocial behavior may once, in its phylogeny, have been altricial: The red-spot-on-parent's-bill pecking may have been selected for, and then the hatchlings may have become more fond of red berries than regurgitated worms. Not pecked at, the parents had less and less occasion to engage in active rearing of their offspring. Not reared, the offspring in turn did little rearing. Life goes on, indifferent to the zoological classifications. Not all phylogenetic developments of behavior need have anything directly to do with evolutionary changes in the species exhibiting the developments.

Dewey makes much of the "altricial" behavior of featherless bipeds, the helplessness of human beings at birth. The intuitive appeal of this strategy is hard to deny. But the lesson of the foregoing examples is this: There may be nothing inherently altricial as opposed to precocial about the feeding behaviors of infants.[61*] Moreover, once we look beyond sucking to grasping and other behaviors, the degree of original organization of human behavior seems all the more a function of the environment with which the infant interacts.

3.13 The Infant's Need for Social Commerce

Dewey argues for the primacy of habits over impulses, despite the latter's temporal priority:

> In the life of the individual, instinctive activity comes first. But an individual begins life as a baby, and babies are dependent beings. Their activities could continue at most for only a few hours were it not for the presence and aid of adults with their formed habits. And babies owe to adults more than procreation, more than the continued food and protection which preserve life. They owe adults the opportunity to express their native activities in ways which have meaning. Even if by some miracle original activity could continue without assistance from the organized skill and art of adults, it would not amount to anything. It would be mere sound and fury.[62]

There are two partly distinguishable claims being made concerning the need for infants to have commerce with adults. The first connects that need with the issue of physical survival, the second with the issue of the ontogenesis of meaning. With respect to the first claim, Dewey's position seems to be that the unorganized impulses of infants can serve vital interests only insofar as adult caretakers take care to manipulate the environment to that end. Thus, the sucking of the infant needs the mother's proper positioning of her breast in order to become vital suckling. As the world is, Dewey's claim holds true, but his suggestion that the world would have to be miraculously different in order for those impulses to be initially efficacious apart from adult agency is a slight overstatement. It is empirically possible, however improbable, that the world once satisfied hungry infants without adult intervention—not by magic or miracles but by virtue of the newborn's physical proximity to (say) coconuts with porous, nipplelike protuberances.

Apropos of his second claim, Dewey contends that "the *meaning* of [the infant's] native activities . . . depends upon interaction with a mature social medium."[63] Adults, by reacting as they do, from habit, to infantile reactions (e.g., crying) convert them into meaningful phenomena (e.g., crying for food). And in order for the baby's activities to reach a "mental" level, to involve some awareness by the baby of their meanings, the baby further requires "that organized interaction with other living creatures which is language, communication."[64]

Helpless babes enter a world, a physical environment, that does not immediately satisfy all their vital needs.[65*] More mature human organisms intervene and complete for the infant its purely impulsive activities, which are not yet organized by coherent aims. Thus, seeing her infant grasping and sucking, a mother may offer her breast (or a baby bottle). Or, responding to the evident distress of the crying infant, a parent or other caretaker may try to satisfy whatever needs might have prompted the distress, or more generally may try to modify the environment to eliminate the crying while ensuring the well-being of the infant. As the caretaker's habits of caretaking intersect with the impulsive activities of the infant, those activities become habituated, progressively integrated with a physical and social environment organized to satisfy the infant's needs. Conceding some sentience to the infant, we may suppose that the effect of the environmental organization upon the infant's experience will be to change it in untold subjective ways, but not yet to create in the infant consciousness of that experience's meaning—apprehension, for example, of distress as need.

There is, sooner or later, a breakdown: The infant's dynamic system loses its equilibrium. Consciousness emerges in the vicinity of the disruption and is directed upon the situation as problematic. The infant's organization of habits does the directing as felt need comes for the first time to have a conscious meaning, a personal significance, for the infant.

Since all habits may be construed as solutions to problems, it seems paradoxical to suggest that the infant's habits precede its own problems. But these infantile habits are solutions to organic problems of which the infant is not yet aware, and these solutions were supplied not by the infant but by its caretakers, who have channeled its impulsive activities along need-satisfying lines. Once those solutions afford a dynamic network of organized habits, the infant becomes capable of an awareness of organically based problems as its own. Until this actually happens, it is a mistake to speak of the infant's *personal* problems, wants, or needs, for persons are, according to Dewey, their habits.

"Distance receptors" of sights and sounds play an important role in Dewey's account of complex animal behaviors: By bringing a larger portion of the organism's physical surroundings within the field of experience, these receptors (together with locomotor organs) allow for a spatially and temporally extended field of activity—a field in which a "response toward what is distant is in effect an expectation or prediction of a later contact."[66] Such expectations are wrought by habit, by the enduring experience of prior activities that have terminated in contact activities. Echoing Craig's distinction between "appetites or aversions" and "consummatory acts," Dewey calls the habituated responses toward distant items of experience "preparatory or anticipatory activities" and the subsequent contact-responses "fulfilling or consummatory activities." Then, apropos of the subjective side of the habituation process, he suggests that "each immediate preparatory response is suffused with the consummatory tone of, for example, the sex or security to which it contributes."[67] Higher brutes and human beings are both subject to this suffusion phenomenon, but there is a difference in the case of the brutes: "Complex and active animals *have* . . . feelings which vary abundantly in quality, corresponding to distinctive directions and phases—initiating, mediating, fulfilling or frustrating—of activities, bound up in distinctive connections with environmental affairs. They *have* them, but they do not know they have them."[68]

There are in nature, Dewey suggests, three distinguishable plateaus of, fields of, interactions. Brutes are beyond the field of purely physical interactions that is the subject of physics; but brutes have not achieved the level of "mind as intellect; possession of and response to meanings."[69] Brutes are on the psychophysical plateau, the level of life, which includes plants and animals. Unlike plants, brutes have sentience; but their (quasi-)purposive behavior, unlike that of humans, lacks conscious meaning.

Since Dewey concedes so much to higher brutes, it is difficult to see just what he denies of them: Their organic activities are arranged (nay, genetically prearranged) into preparatory-consummatory series that are at least potentially significant. The brutes are moved to action by their feelings, which are "suffused with consummatory tone," yet are not aware of what those feelings mean. The activities of higher brutes are subject to habituation but not ("so far as we can judge") to what Dewey calls "mediation": *"The expression of every impulse stimulates other experiences and these react into the original impulse and modify it. . . .* This back-reference of an experience to the impulse which induces it, we may term the *mediation* of impulse. . . . It is evident that . . . mediations . . . constitute the *meaning* of the impulse—they are its *significance,* its *import.* The impulse is *idealized."*[70*]

Higher brutes would seem to lack ideas, the conscious meanings of acts. Ideas make it possible for human beings to deliberate.

3.14 The Mystery of Deliberation

The need for and nature of what Dewey calls "deliberation" might be illustrated by returning to the earlier analogy of a spider faced with the task of reconstructing its web. The web's breaking stirs the impassive spider to consciousness. But even if the vibrations of its web direct it to the disaster area, the perplexed spider requires further guidance for the task of reconstruction. How does the spider gain sufficient distance and autonomy from the damaged web to be guided by it intelligently? Well, perhaps the spider does not always have to gain this. Thus, it might back up along the broken strand (a habit reified) and start running toward the break, thereby guided by its past achievement. Taking a running leap from the point at which the strand is broken, the spider might spin out a silky substance of new strand in its own wake, along the trajectory determined by the old strand. The spider might begin again, spin again, after the pattern of its prior accomplishments, without ever exactly gaining a larger perspective upon its activity. Insofar as old strands merely need to be extended, this strategy is equal to the task; but if the old strand has been a mistake, having been attached to an unstable, unreliable environmental support, then, using this strategy, the spider is fated to repeat the error.

At some point, perhaps, the *habit* of reconstruction stands in need of reconstruction, too. Calling the strands *first-order* habits and the habit of reconstructing them *second-order* habits, it might be said that the failure of second-order habits lies in the failure to reestablish first-order habits. Thus, when the spider takes a running leap from the breach point of its old strand, the force of momentum may sometimes catapult the spider not to a slightly more distant branch or twig, but into the void. Swinging helplessly on the end of its unreattached strand, the spider might be said to be awakened from its reconstructive complacency and to be made to realize the need for a new mode of reconstruction. From its new vantage point, the spider might gain further perspective on its own strands (habits) and their possible and actual environmental supports.

But a better visual perspective is not enough: The spider needs to be able to reason effectively about what it sees. The spider, as rebuilder in a changed environment, needs to be able to think, however roughly, the following kinds of things: "Well, the twig is gone. What can I do? Maybe if I head off more to the left, I can reach that other branch; maybe I can simply reinforce the strands on both sides of the broken ones," and so on.

One might suggest, borrowing a thought from Kant, that the myopic spider needs conceptual "eyeglasses," to ensure that its percepts will not be, for its intents and purposes, blind. But the best-fitting concepts in the world will not make intelligent foresight possible unless they can somehow be used to see beyond the present situation to what is absent to any observer, however farsighted. What the spider needs is something that, on an intellectually clear day, enables it to see into the possible future.

Köhler subscribes to a familiar myth that the spider (which stands for us) has an inner-inner life in an ethereal world modeled on the one external to it. Although it may have to be initiated by perceptual processes channeled by conceptual eyeglasses, this innermost world is presumed to have a somewhat autonomous existence; and the spider can live its own inner-inner life within this world, now subject only to the changes wrought by the spider's own will, no longer subject to those natural laws that govern the world beyond. (Variants of this myth differ concerning the locus of those laws but agree that they are not binding on the innermost world, unless the spider chooses them to be. Other, psychological laws may of course determine its choices.) This inner-inner world, once removed from the inner world of perceptual representations, is where Köhler would have us conduct our ideal experiments, our deliberations about the future course of events. But Dewey denies the existence of this particular world, so there is no such place for his deliberations to work out problems, in solitude, before attempting a reconciliation with the natural world.

Dewey's solution to the mystery of deliberation is to propose a special role for language (as "conversation, whether it be public discourse or that preliminary discourse termed thinking"). This role demands no artificial stage, no inner-inner world, in order to be played effectively:

Brute efficiencies and inarticulate consummations as soon as they can be spoken of are liberated from local and accidental contexts, and are eager for naturalization in any non-insulated, communicating, part of the world. Events when they are named lead an independent and double life. In addition to their original existence, they are subject to ideal experimentation: their meanings may be infinitely combined and re-arranged in imagination, and the outcome of this inner experimentation—which is thought—may issue forth in interaction with crude or raw events.[71]

It is language that sets meanings free from the activities in which they originate and *empossibles* (to coin a word)[72*] deliberation in signs. The sign vehicles, whether inner or outer, whether private images or public utterances, allow for the "vicarious presence in a new medium"[73] of possible future events. By consciously manipulating these signs, one gains the requisite perspective for that intelligent reorganization of habits that Dewey calls deliberation; for in thinking with signs, it is possible to think about activities without actually engaging in them.

3.15 Crossed Purposes: Dewey's Concession to Instinct

Dewey's fully developed account of deliberation would seem to go far beyond any theoretical reliance upon the notion of instinct. And if one were to look back at the less advanced, predeliberative stages of functioning that are posited by his theory, one would still discern among them some clear alternatives to instinct. Thus, the fairly rigid, routine habits and the more flexible, intelligent habits are both acquired rather than innate. But if one looks even further back, to the original impulses that give rise to habits, one finds that although those impulses lack the vital organization that might qualify them as instinctive, they do have instinctual origins.

Dewey, following James, regards human beings not as devoid of instincts, but rather as having a superabundance of instincts at cross purposes: "Man can progress as beasts cannot, precisely because he has so many 'instincts' that they cut across one another, so that most serviceable actions must be *learned*."[74] Dewey might here be taken to suggest that impulses are *unorganized*—that is, uncoordinated with vital environmental ends—in consequence of conflicts among instincts underlying those impulses. Contrary varieties of instinctive organization effectively cancel out one another, leaving behind what appear to be nothing more than "loose, undirected"[75] impulses.

This suggestion, though not inconsistent with his theory, is a little difficult to reconcile with other tenets of Dewey's theorizing: Subsequent, acquired organizations (of human tendencies to act) are never entirely wiped out; even when they are disorganized (i.e., when there is a breakdown of habits), those subsequent organizations endure, serving as materials for deliberation, as the stuff of later reorganization. How curious, then, that our later acquisitions should prove so much more permanent than our original, biologically inherited patterns of behavior—after all, the received opinion is that it is in the nature of instincts not to be easily gotten rid of.

James's view does not, as Dewey's version of it seems to, imply that instinctive organization is dispatched by crossed purposes. James says that "however uncertain man's reactions upon his environment may sometimes seem in comparison with those of lower creatures, the uncertainty is probably not due to their possession of any principle of action which he lacks." [76] And although he attributes the uncertainty of our reactions to the fact that our contrary instincts are so prevalent that "they block each other's path," [77] James does not commit himself to any resultant disorganization of those instincts: Blocked from acting, they simply wait their turns—" 'experience' in each particular opportunity of application usually deciding the issue" [78] of who goes first.

Dewey and James appear to have proposed two distinct empirical possibilities. How might one decide, on empirical grounds, between them? If James's view is correct, then the infant's behavior will be as changeable as the weather: Sometimes the infant will do one thing; sometimes, when a contrary instinct prevails, quite another. One might expect that subtle differences in the situations eliciting divergent reactions would indicate, to the acute observer, that different instincts were in operation; but if human instincts truly "*contradict each other*," [79] as James says they do, then sometimes even exactly the same external circumstances will, depending upon internal factors (say, a momentarily greater strength of one impulse over its rivals), produce contrary courses of reaction. Such inconstancy would confound one's efforts to study the matter inductively: Even if James's view were correct, it would be difficult (indeed, still more difficult than ever) to see how any observational or experimental scrutiny of the infant's overt behavior could serve to confirm the existence of the (conflicting) instincts posited.

A similar difficulty attends Dewey's view, assuming it to be correct: Lacking any coherent principles to guide them, the infant's impulsive behaviors will not be uniform responses to particular environmental stimuli; the responses will only later be regularized, by the caretaker's efforts to meet the infant's needs. The failure to detect any instinctive organization of the infant's impulses is certainly no embarrassment for Dewey, but the fact that the infant's observed behaviors might be indistinguishable from those that would be expected were James's view correct is surely no point in favor of the Deweyan alternative, either.

Since both views presume a continuity of sorts between human and other animal behavior, it might be thought that the appropriate strategy of empirical confirmation for the views would be a less inductive, more theory-laden phylogenetic line of argument on their behalfs. And since behavior leaves no fossil records, since evolutionary development among higher organisms cannot be duplicated in the laboratory (the way some such developments of lower life forms, which multiply far more rapidly, can), and since our phylogenetic ancestors are inaccessibly extinct, it seems we can only thus argue from a contemporary species-comparative standpoint. Were either of the views correct, we might expect to find some evidence that manifest, nonconflicting instincts have become less prevalent among organisms closer to our level of development, where levels are gauged by a sense of which (higher to lower) species approximate our (recent to remote) phylogenetic ancestors. But even assuming that something like the deprivation experiment could yield a reliable inventory of (nonconflicting) instincts for these phylogenetic relatives, the findings might not unequivocally support the common thread of James's and Dewey's hypotheses. Thus, a declining number of manifest instincts associated with increasing closeness of developmental level, would be compatible with the contrary hypothesis that instincts are decreasing in number (rather than increasing but contravening one another). And as a basis for deciding between James's and Dewey's alternatives, the phylogenetic evidence again seems uselessly consonant with both.

Ignoring the problem of confirmation, one might still ask what sort of phylogenetic story would make sense of either of the two alternatives. How, after all, could conflicting instincts evolve? One fairly plausible answer is this: Each instinct covers a range of possible environmental circumstances; each may be elicited by many distinct stimuli. New instincts that handle circumstances outside the range of those covered by old instincts sometimes emerge. But these new instincts also cover a range of circumstances, some of which are already covered by the old instincts. Such overlapping may eventually be complete, when enough new instincts evolve. When this happens, *Ecce Homo sapiens*.

But another story might be proposed to fit the contrary hypothesis that instincts are actually decreasing in number: Instincts are *improved* by phylogenetic developments. As instincts become increasingly flexible, increasingly intelligent, they become less akin to their phylogenetic prototypes. Yet another story might suggest that intelligence evolves independently of instincts, whereupon nature favors their elimination, and intelligence emerges victorious.

It is risky business to argue from several cases, eliminating all but one as implausible, unless one has good grounds for eliminating them and reasonable assurance that all the possible cases, all the empirically sensible stories, have been given. Still, I would venture to say that the least plausible story is the last one given. Its principal shortcoming is that the independent evolution of intelligence is all too conveniently assumed without any real indication of how it could take place. Summarily dismissing that story, I would further observe that the remaining candidates share a notable feature: They posit some flexibility of instinct, some range of stimuli capable of eliciting given instinctive responses. This position is not implausible but does raise some question about the true identity of the characters in both stories: Are they really instincts?

In sum, Dewey's apparent concession to instincts may not be incoherent, but it has no special merit, either: Both James's prior version of that concession and the contrary claim that humans altogether lack instincts are also compatible with the same range of inductive findings and are, phylogenetically speaking, equally sensible suggestions. Moreover, all three views would seem to share a certain sympathy for the idea that behaviors are flexible, an idea that on its face seems antithetical to the notion that behaviors are instinctive. This superficial impression merits fuller investigation, but a useful preliminary will be to consider one more dynamic theory of behavior, one that goes still further away from any commitment to instincts.

Chapter 4

The Epigenesis of Behavior

Zing-Yang Kuo's epigenetic theory bears many striking similarities to Dewey's theory but manages to avoid the one unfortunate concession that Dewey, after James, makes to the notion of instinct. Even though Kuo's theory is rather tentative and supported by sketchy, merely suggestive empirical findings, it can nonetheless be said to afford the strongest, theoretically most compelling and unified scientific case yet to be made against instinct. This chapter critically examines, by turns, the basic concepts of Kuo's theory, some philosophical-methodological issues raised by it, and selected problems and prospects for its associated research program, as contrasted most specifically with that of Lorenz.

4.1 Kuo's Epigenetic Standpoint

Zing-Yang Kuo conceives his programmatic theory of behavior to be a revised version of the "most radical Watsonian behaviorism" (to which Kuo's early writings almost notoriously subscribed); but one might equally well construe Kuo's theory as a natural extension of some Deweyan themes, including those Dewey relies upon to dispute the instinctive character of vital organic activities: Dewey insists that the whole, ever-changing organism is involved in each of its acts; that the environment on which they impinge is continuously varied; and, hence, that the fit between the acts and their environment is far too variable to be preestablished innately. But the organism about which he articulates these holistic-flux themes is already an infant, and Dewey seems content to view embryological development as little more than pure maturation, terminating in a superabundance of conflicting instincts that cancel out one another. Kuo, on the other hand, extends those themes to the case of prenatal activities and the intrauterine environments in which they occur.

Gilbert Gottlieb sums up some major tenets of Kuo's theory: "According to Kuo, prenatal (as well as postnatal) behavior is a function of three simultaneously interacting factors: (1) the embryo's [or organism's] developmental history, (2) the present (or immediate) environmental context, and (3) the physiological condition of the embryo." [1] The similarity to Dewey is especially evident in connection with the last two factors, since Kuo's notion of the *environmental context* "stresses the situational or 'field' determinant of the organism's immediate behavior," and his notion of the organism's *physiological condition* stresses that the whole organism, including its "metabolic biochemical and other (typically unobserved) interior events," determines each of its behaviors.[2*]

The term *epigenesis* commonly signifies a biological theory that embryonic development is a sort of order from chaos, a progressive organization of initially disordered structures. It is not incompatible with such epigenesis to view development as purely maturational. For instance, E. Mayr characterizes epigenetic development as "the *interaction of genetic factors* during the developmental process"[3*] (emphasis mine); yet he also, not inconsistently, characterizes ontogeny as a manifestation "of the decoding of the information embodied in the genotype."[4] So, too, in the case of G. E. Coghill's theory, which might be thought the very paradigm of an epigenetic view: Coghill is known for his intriguing idea that prenatal behavior begins with a "total pattern" that is progressively differentiated into more "partial patterns" or "local reflexes."[5*] But he considers this course of development to be a function of the growth of the nervous system; and he regards behavior as a mere "epiphenomenon of the nervous system."[6]

The special character of Kuo's epigenetic viewpoint is explained by Gottlieb in terms of a distinction between *predetermined* and *probabilistic* views of epigenesis: "One viewpoint holds that behavioral epigenesis is predetermined by invariant organic factors of growth and differentiation (particularly neural maturation), and the other main viewpoint holds that the sequence and outcome of prenatal behavior is probabilistically determined by the critical operation of various endogenous and exogenous stimulative events."[7] Kuo's view is said to be one of probabilistic epigenesis, which Gottlieb takes to entail a *bidirectionality* thesis: "The assumption of reciprocal effects in the relationship between structure and function whereby function (exposure to stimulation and/or movement or musculoskeletal activity) can significantly modify the development of the peripheral and central structures that are involved in these events."[8]

E. B. Holt, also deemed a probabilistic epigeneticist, suggests a mechanism whereby such bidirectional effects might occur—a neurodevelopmental mechanism that explains how neural networks are established, how an axon of one nerve cell comes into contact (forms a synapse) with a particular dendrite of another nerve cell. This is said to take place because of electrical potentials created by stimulation: They guide the growth of dendrites toward loci of "greatest neighboring stimulation."[9*] Thus, the embryonic nervous system quite literally grows to the way it has been exercised, stimulated, by behavior.

But Kuo resists such efforts to explain the bidirectionality of structure and function in terms of specific neural mechanisms, for some of the same reasons that Dewey resisted James's neurological account of habit: Such explanations are far too reductionistic and too ready to take what might even be part of the truth for the whole truth. Kuo's theory, like Dewey's, is wedded to the idea that larger organic contexts than the central nervous system (CNS) must be considered in trying to explain behavior.

4.2 The Concept and Theory of Behavioral Gradients

Kuo expresses the idea that behavior involves a large organic context by proposing a concept of *behavioral gradients*. This concept is intended to replace such older, inadequate conceptualizations of behavior as "total pattern" and "local reflex." What these older concepts would seem to ignore is that "behavior is not merely motor movement or glandular secretions; it includes activities of every part and every organ of the whole animal as well as their feedbacks and interactions." [10] Kuo characterizes his proposed replacement as follows: "At every level, from metabolic changes in the tissues to overt bodily movement, there are both quali-tiative and quantitative variations from moment to moment. *These variations are what we mean by behavioral gradients.*" [11] The connection between this concept and the methodology of "radical Watsonian behaviorism" would seem to reside in the idea that, although some gradients or variations are visible ("explicit") while others are invisible ("implicit"), "visibility is a matter of degree," [12] and all gradients are subject, in principle, to objective observations (using suitable observational instruments if not the naked eye).

Substantive theses belonging to Kuo's theory of behavioral gradients include the following: 1) The variations and their feedbacks (which are, presumably, fur-ther variations) "are formed into complex, interwoven but definite and orderly . . . *patterns of behavioral gradients.*" 2) "These patterns are in a process of continu-ous change, for the different parts and organs vary constantly in intensity and extensity." [13] (This seems to be why Gottlieb says, with Kuo's likely approval, that the conception of behavioral gradients is intended "to call attention to the fact that the [organism] is more or less active all the time." [14] 3) "The goal of the science of behavior [is] to discover the ordering and laws of such chang-ing patterns of gradients." 4) "The activities or changes taking place inside the body are not just physiological bases or physiological counterparts of behavior; they are inseparable portions of the total pattern of behavioral gradients," the "integrated, inseparable total pattern in which the activity of the different parts and organs varies from time to time in intensity and extensity." [15] Summing up these points, Kuo remarks: "*The most essential feature of the behavioral gra-dients concept is that, in any given response of the animal to its environment— internal or external—and in any given stage of development, the whole organism is involved.*" [16]

On this theoretical foundation, Kuo argues against any further reliance upon the concept of a local reflex or any other concept that presumes the existence of independent units of behavioral response to stimulation: However local a given response may seem, it changes the overall "organization of the gradient system at [its current] stage of development." [17] (Of course since we cannot specify in advance of details forthcoming only after Kuo's theory begins to pay off the precise sense in which any given reaction involves the whole "gradient system," we are not in a position to discount the possibility that sufficiently isolated changes in the system will prove compatible with the hypothesis of [relatively] local reflexes. Accordingly, we cannot even say that Kuo's theory would, if true, vitiate the concept of a local reflex, but only that the developed theory might do so.)

Kuo is prepared to retain "as a convenient abstraction for descriptive purposes" such notions as that of "fixation of habit," but he warns against viewing broad patterns of overt behavior as "fixed actions" or "stereotyped behavior." His point is that such construals "overlook or ignore the constant variations of the gradients." [18] Like Heraclitus's river, the organism's gradient patterns are never twice the same.

4.3 The Concept and Theory of Behavioral Potentials

Combined with his concept and theory of behavioral gradients is Kuo's concept and theory of *behavioral potentials*. His rather informal characterization of the concept is this: "By behavioral potentials we mean the enormous possibilities or potentialities of behavior patterns that each neonate possesses within the limits or range of the normal morphological structure of its species." [19] But this concept is not meant to apply only to the neonate: Throughout the life of the organism, it possesses a range of potential behaviors, a range that varies continually. Kuo compares talk of behavioral potentials of a neonate to talk of the "utilitarian potentials" of a piece of wood from a tree: "The log can be used as firewood, molded into a part of furniture, burnt into charcoal, ground into pulp for paper, buried underground and after millions of years turned into coal, carved into a statue, or innumerable other possibilities. By the same token; a newborn kitten has no genetic factor that makes it a rat killer, but it does possess the *potentialities* for becoming both a rat killer and a rat lover." [20]

Kuo's rather ordinary talk of potentials carries an extraordinary message about the plasticity of behavior, the range of patterns it may exhibit. This talk also gives new meaning to the old saw "a chip off the old block"; for even identical twins, two chips off the same block, will differ in their behavioral potentials: "Here we must also stress the fact that no two pieces of wood from the same tree have exactly the same texture and appearance; thus, their utilitarian potentials are different. The phenotypes of the two pieces of wood are different, but their genotype is the same. Just as no two pieces of timber from the same tree are identical, no two persons are born with precisely the same behavioral potentials." [21]

Kuo's concept of behavioral potentials serves to clarify his sense of the *undeni-*

able residuum of hereditary determination in the ontogenesis of behavior. Kuo says, "The epigeneticist does not repudiate heredity as such, but he maintains that genetic factors merely set the boundary to the potential range, which is far broader than the actual behavioral repertoire that an organism realizes during its development." [22] Yet Kuo wants to deny that behavioral potentials might be identified with "inborn behavioral predispositions or behavioral tendencies." [23] The realities of behavior are, for Kuo, the behavioral gradients and their patterns. Behavioral potentials are not *real*, in any proper metaphysical sense of the word; they are nothing more than "the possibilities of behavior patterns." [24] Organisms have these potentials, but they can no more be said to be *in* the organisms than a furniture part can be said to be in the log from which it could be made.

The functional variables of Kuo's prospective epigenetic theory are what he calls "the determining factors of behavior." Kuo groups these factors "into five main categories: (a) morphological, (b) biophysical and biochemical (physiological), (c) developmental history, (d) stimulus or stimulating objects, and (e) environmental history." Since "behavior is a functional product of the dynamic interrelationship of these five groups of determining factors," [25] it might be said that the range of behavior potentials is the range of these functional products, each of which is a particular pattern of behavioral gradients. Accordingly, it might be reemphasized that Kuo has no "ontic commitment" to behavioral potentials: They are not among the values of the variables of his theory.[26*]

4.4 The Diminution of Plasticity Thesis

The fact that behavioral potentials are not among the posits of Kuo's theory does not prevent him from making some substantive theoretic claims about the ranges of those potentials. Gottlieb mentions two particularly important claims on this topic. The first is that the "developmental history of the individual is the major (though not the sole) limiting factor on the range of behavioral potentials." [27] Kuo recounts a personal anecdote with some such claim as its analogical upshot: He once asked a Chinese Buddhist scholar and fortune-teller to tell him (Kuo) his fortune and was told in reply, "If you tell me what you have been, I'll be able to foretell what you will be." [28]

Kuo's claim may be seen as his reminder to other theorists who have tended to ignore the importance of developmental history. Thus, Kuo charges that "Pavlov . . . was . . . unaware of the fact that the bond between the so-called unconditioned stimulus (food) and the unconditioned reflex (salivary secretion) was formed by the conditioning process early in the life of the dog." [29] And Lorenz, in consequence of like neglect, is said to misidentify various animal behaviors as species specific: Kuo contends that although the European chow dogs Lorenz once studied may have had their fighting "almost always end harmlessly" with the weaker dog's throat-exposing submission to the stronger dog, this is hardly the "remarkable product of adaptive behavior through evolution and natural selection" that Lorenz takes it to be.[30] For in cases where Chinese chow dogs—raised by Chinese peasants who allow them, as pups, to fight "neighbors' dogs to the finish"—respond thus submissively, the "response is an invitation to certain death."[31] Unconditioned responses and species-specific behaviors are prime candidates for innate forms of behavior, and Kuo's emphasis on the developmental-historical determinants of these candidates calls their qualifications for such office into serious doubt.

Kuo's second major claim about ranges of potentials is "that successive stages of development are correlated with a diminished plasticity or narrowing of the range of potentialities."[32] And this second claim may help to explain the first: Developmental history may be the chief delimiter because it is a progressively greater delimiter of behavioral potentials. Thus, the longer one has lived, the more one has to tell the Buddhist fortune-teller; and the more he is told, the better able he is to foretell one's future behavior.

Without stopping to gauge the strength of this connection between claims, let us try to take the measure of the second claim, Kuo's diminution of plasticity thesis. Were one's life possibilities to be set out, schematically, they might be signified by a vast, treelike structure of lines, starting out in one direction but quickly branching out in many others. Were development viewed as movement along these lines, any paths taken would delimit one's potentials to possible paths ahead, effectively diminishing the size of the original array of paths available at the onset of ontogenesis.

Now we may properly lament some paths not taken and truly observe that they would have afforded more alternatives. We may even be saddened by that existential plight occasioned by the fact that we can follow only one path; but we ought to remember that same fact when we are tempted, wrongly, to suppose that our actual path is inevitably a substantive diminution of potentialities, a progressive loss of plasticity. No path we actually take will have as many branches as the whole tree, but the whole tree was not a path that could have been taken—nor are all the remaining branches before us. We may lose plasticity in virtue of some actual paths we take, but not merely because we have had to take an actual path instead of the whole tree. The whole tree never was an option, sad as that may seem or be.

As a substantive theoretic claim, Kuo's diminution of plasticity thesis cannot be derived simply from the concept of a behavioral potential, from the abstract, treelike structure of an individual's range of behavioral potentials. Such a tree might, for all we know, involve some repetitions of alternatives along possible routes. If so, our earlier failure to exercise a given option might not deprive us of it—notwithstanding the fact that our historical development on whole would not be identical to what it would have been had we taken that same option the first time.

If support is to be found for Kuo's diminution thesis, then it will be necessary to look among other, more substantive details of his epigenetic theory. One promising place to look would seem to be Kuo's view of the reciprocity between structures and functions, his version of the bidirectionality thesis. One might, at any rate, attempt to restate the diminution thesis in terms of this other thesis, thus: functions, throughout the course of development, give rise to modified forms, both morphological and physiological; these modified forms in turn determine (in conjunction with environmental factors) behaviors within a progressively narrower range of behavioral potentials.

But Kuo might resist the emphasis given in this version of his diminution thesis to morphological and physiological forms; for he contends that "the combined factors of developmental history and environmental context alone are often sufficient to reduce the range of behavioral potentials, a reduction that does not necessarily involve anatomico-physiological factors; it is a reduction of plasticity in the formation of new patterns without any need for reference to mythical predetermined neural organization." [33]

Critics might suppose that Kuo's radical behaviorism had here got the better of him, leading him implausibly to allow the forms or structures responsible for diminished plasticity to float homelessly in the air. But Kuo has available this Deweyan defense: The forms do not exactly lack citizenship in the morphological-physiological system; they are citizens of the larger world of nature. The whole organism is involved in each of its acts, so the forms in question are not unrelated to the organism's morphology and physiology; it is just that those forms reside principally in the environmental context and the developmental history of the organism.

4.5 Epigenesis as Historicism:
Laws and Other Patterns of Development

What may continue to puzzle his critics is Kuo's apparent willingness to regard developmental factors as distinct variables, independent of any morphological and physiological factors. It is one thing to observe that Kuo, like Dewey, lacks reductionistic tendencies of thought, but quite another to make sense of a suggestion that seems to reify developmental histories. Kuo's theory is intended to apply to individual organisms and predict their future behavioral patterns not just on the basis of present patterns, but also on the basis of past patterns and their sequences. And Kuo's talk of prophecies based upon developmental histories further suggests that he construes such sequences of past patterns as *trends*.

Karl Popper has roundly condemned at least some positing of trends as a methodologically incoherent species of what he calls "historicism"—namely, "an approach to the social sciences which assumes that *historical prediction* is their principal aim, and which assumes that this aim is attainable by discovering the 'rhythms' or the 'patterns,' the 'laws' or the 'trends' that underlie the evolution of history." [34] But Popper's main quarrel is with the idea of "*absolute trends;* trends which, like laws, do not depend upon initial conditions, and which carry us irresistibly in a certain direction into the future[—trends which] are the basis of unconditional *prophecies,* as opposed to conditional scientific *predictions.*" He claims to have no quarrel with "those who see that trends depend on conditions, and who try to find these conditions and to formulate them explicitly." [35]

It is moot whether Popper has a quarrel with the major historicists he condemns, since even Marx lists *classes* among the conditions under which certain trends (of class conflict) will persist. And Kuo's trends are surely not absolute, either, so Popper could not on that ground condemn them as untestable: Not only do they depend on empirical conditions (viz., past gradient patterns); Kuoian trends would seem all too close to being constituted exclusively of those conditions.

But other objections press in: If past patterns are to affect future patterns, then those past patterns must be around to do so—otherwise their causal efficacy is a mysterious, scientifically anomalous sort of action at a distance in time. Moreover, standard logical methodology of scientific explanation would expect Kuo's historicistic trends to be lawlike tendencies that, together with the present state of the organism and its environment, sufficed to explain, to predict, the subsequent behavior of that organism. But Kuo's trends are idiosyncratic to individual organisms, so it might be objected that the trends lack the generality associated with genuinely scientific laws and so fail to satisfy the methodological criteria of scientific inquiry.

One response Kuo might give to the first objection is that developmental histories operate by way of their "organic trace effects" [36] on morphology and/or physiology and so are temporally connected to the behaviors they help to determine. A stock rejoinder to the second objection would be to say that such trends were only de facto idiosyncratic: They would still be logically or methodologically general, in that they would apply to any organism that, pace Kuo, shared the individual's genotype and exact developmental history. But this rejoinder raises a further question: How does one ever discover a singular law or trend? The short unsympathetic answer is that qua scientist, you never do, but that qua behavioristically inclined fortune-teller, you can at least approximate such a discovery. Given that laws of science are regularities of nature, whatever else they might be, this negative reply is tempting, for we discover regularities by dint of their recurrent manifestation. But even in the absence of any recurrence, some larger theory might dictate the existence of such laws operating behind the scenes; and these laws might then be confirmed, indirectly, by way of empirical evidence for the larger theory to which they contribute.

Suppose something like the following takes place: Fairly general laws that apply to a host of different organisms (say, all vertebrates) are discovered. These laws are combined with other, even more general laws of animal behavior. The result is a theory still inadequate to the task of explaining and predicting the behavior of any particular vertebrate animal. Then someone notices a sort of pattern to a given individual's developmental history, a pattern of behavioral gradients that organizes them "in space and in series."[37] Combining details of this pattern with the inadequate general theory, we become able to predict very successfully the animal's behavior. But even after discovering patterns for an enormous number of individuals, we never do manage to find lawful regularities among any of the patterns: For one thing, elements of these patterns are temporally discontinuous, so the patterns themselves are not lawlike conjunctions of events; for another thing, we are unable to correlate these temporally extended patterns with any current physiological-morphological factors (which we might hope to use to account for the apparent causal efficacy of the patterns themselves).

What should an intelligent onlooker make of the (hypothetical) fact that we can use our hodgepodge theory, composed of laws and patterns, to predict what vertebrates will do in particular circumstances? A few successful predictions might well be dismissed as some lucky guesses by fortune-tellers in scientists' garb; but if the patterns could reliably be discerned and consistently used to make correct predictions, it would be silly to suppose that luck had simply become phenomenal. Rather, fortune-telling would have been turned into a science.

4.6 Habituation and Diminution:
Deweyan Prospects for Kuo's Theory

Returning to the diminution of plasticity thesis, we should note that diminution would appear to have something to do with the progressive patterning of the field of behavioral gradients, that at least some patterning would seem, for Kuo, to amount to a condition of decreased plasticity. It is also fairly plain that, since plasticity is said to decrease with every stage (or, better, point) of development, Kuo presumes the patterning that accounts for diminution to extend back over the entire developmental history of the organism. And given Kuo's view that the whole organism is involved in each of its acts, we may even suspect that all patterning is total, that it organizes all behavioral gradients past and present. But this strongly suggests that there is a particular pattern for any individual organism and, hence, that all ontogenetic patterning amounts to a diminution of behavioral potentials. If this is Kuo's thesis, then it begins to look as though he has misapprehended, in ways I earlier cautioned against, the organism's treelike array of behavioral potentials.

Dewey, by comparison, would seem to have the far more plausible view that only some patterning—namely, that which is associated with blind, utterly routine habits—might be said to diminish our behavioral plasticity. Yet why should his theoretical outlook, which shares Kuo's holism, give Dewey any clear advantage? He does posit partial patternings, individual habits, but he also views them as internally related to the whole spread of the dynamic system, to the total network of all the organism's behavioral tendencies. Some such scheme is, given minor adjustments in theory, also available to Kuo.

Kuo does not even seem willing to countenance *individual acts,* which might be thought to be the telling effects of individual habits or tendencies; so let us consider, first, how such acts might be individuated against an epigenetic background, against an organism's total pattern of behavioral gradients. The wherewithal for this individuation is provided by Kuo's characterization of "behavioral gradients as differential intensities and extensities of involvement of different body parts or organs of the whole organism in its response to environmental stimulation." [38] Given this characterization, it might be said, for example, that what we ordinarily refer to as a chick's individual act of pecking is a partial pattern of maximal intensity and extensity of involvement of neck, eyes, beak, and other parts in the total pattern of behavioral gradients over specific periods of time. The maximal intensity and extensity of involvements of these parts is gauged by way of comparison with the rest of the current state of the gradient system. Where no such involvements clearly stand out, the inevitably active organism may be said, for the sake of convenience, to be at rest; though there should be no suggestion that quiescence is a state of equilibrium toward which the dynamic system tends, for Kuo's theory is one of dynamics without equilibria. All changes of behavioral patterns are said to be "a result of the continuous dynamic exchange of energy between the developing organism and its environment, endogenous and exogenous," [39] so the maximizing of certain involvements, the onset of individual acts, is a direct consequence of the redistribution of energies within the whole gradient system. Waxing metaphoric, we might propose that shifting energies carve individual acts out of the epigenetic chaos of behavioral gradients.

Kuo may not endorse this proposal, but he does come close to committing himself to learned individual acts and underlying habits:

Ontogenesis is a process of modification, transformation, or reorganization of the existing patterns of behavioral gradients in response to the impact of new environmental stimulation; and in consequence a new spatial and/ or serial pattern of behavior gradients is formed, permanently or temporarily ("learning?") which oftentimes adds to *the inventory of the existing patterns of behavior gradients* previously accumulated during the animal's developmental history.[40] (emphasis mine)

The "existing patterns" mentioned here are plausibly interpreted (given that they belong to an "inventory") as repeatable, as members of the organism's repertoire of behaviors. Since total patterns are never the same twice, these repeatable patterns would have to be partial patterns. My proposal affords a handy way to individuate the acts that have these patterns. And, given that the acts are repeatable, it would not be unreasonable to posit, corresponding to these patterns, some learned tendencies to produce them, that is, some habits. The prospects for making most sense of Kuo's diminution thesis, given these positings, remains to be considered; but they seem, in any case, to be compatible with the holism of his theory.

4.7 Patterning and Reduced Plasticity

Kuo's diminution thesis conveys a murky image of the ontogenesis of behavior as a process of shaping (or patterning) some queer medium. This medium is initially quite plastic, and some of the shapes it takes on are temporary, the medium thus resilient; but the results of successive patterning are cumulative and lead, inexorably, to ever greater rigidity. What seems particularly obscure about this image is the connection it makes between patterning and a loss of plasticity, a diminished range of behavioral potentials. The remarks that follow are efforts to clarify and to assess the strength of that connection.

One presumption that might have disposed Kuo to make such a connection is the idea that new behaviors are constrained, somehow or other, to conform to the patterning of a lifetime. This need to conform might seem to make any new behavior less pliable in its own right, for that need would seem to suggest that one's whole past history works against the possibility of proceeding in novel directions. But does rigid adherence to the past inevitably lead to a rigid response, or a more limited range of potential responses, to the present or future? Much depends upon how the organism conforms to its past: if the organism repeats its past response patterns exactly, then surely its responses will in some sense be rigid. But such *rigidity of strict repetition* would conflict with Kuo's holistic-flux themes and so cannot be used to account for his diminution thesis.

Consider though, apropos of that thesis, some ersatz sage advice: Do not, in the future course of your life, stray any farther from your prior path than is necessary—be true to yourself. These words might seem to counsel some type of repetitional rigidity, but in fact they carry a more holistic message involving the idea that all of one's past patterns combine to establish an overall direction (or path) for future behavior. What is counseled, then, might be called the *rigidity of least deviation* (from such a path). Following the advice requires a sense of what one's path would, if unobstructed, be. Let us suppose that this sense might in principle be acquired by means of a "vector summation" of all the individual movements, in every particular direction, that constitute our history of past behaviors. This would be no mean feat, and one reason we might view our past as weighing us down is the recognition that the lessons of our own histories are exceedingly difficult to draw. Yet the rigors of our efforts to draw guiding lessons from our past should not be confused with future rigidity: Once the lesson is drawn, our future path set, it is no more rigid for having been determined with difficulty. Besides, in the case of organisms bound by their developmental histories along certain paths, the idea of drawing a lesson plays no role. The organism's future path is, on Kuo's view, established by the dynamics of the epigenesis of behavior. In general, this need not be seen as demanding the organism's conscious effort—to the contrary, Kuo sees *consciousness* as but one more old-fashioned psychological concept that the epigenetic viewpoint will someday render superfluous.

Kuo seems more inclined to the view that the sheer complexity of past patterning renders the organism behaviorally rigid, but why should this happen? Prior complexity might turn out to be of little moment to the character of a subsequent response, which might not even be comparably complex: Thus, if the character of the response is determined by an automatic vector summation, then the partial pattern of that one response, an individual act, might be no more complex than the pattern of response of a newborn babe. But suppose, for the sake of argument, that the more complex the organism's past patterning, the more complex the patterns of its subsequent responses. Why should this be tantamount to a greater rigidity of those responses? A more complex response might even be thought to allow for more subtle adaptation to complex environmental exigencies, and adaptiveness would seem to connote plasticity, not rigidity. Of course on the opposite side, it might be thought that a highly complex response is more likely to be fine-tuned to a particular environment and hence less likely to be flexibly adaptive. So maybe *some* kind of inevitably increasing complexity of past patterns inevitably leads to *some* kind of increasing complexity of subsequent responses, which, inevitably, amounts to *some* kind of increasing rigidity of those responses; but, pending further specification of terms, we cannot be clear about whether this is so.

Another option for making sense of Kuo's tie between increased patterning and decreased plasticity is to understand that patterning as consisting of proportionately ever greater numbers of partial patterns for routine habits, each of which leads to a (repetitionally) rigid unit of behavior. But a problem for this option arises from the fact that, given the terms of Kuo's diminution thesis, this variety of increased patterning would have to be a trend common to all the behaving organisms to which the thesis applied. The problem is that unless Kuo could account for such a trend, could state the conditions under which it persisted, he might be accused, more justly than before, of what Popper regards as a pernicious historicism. (Indeed, even if Kuo could state such conditions, they might happen to be inseparable, say, from those for the continued life of the organisms and, so, still not help to render the trend hypothesis testable.) Moreover, even if such a trend could be established to exist, the connection with rigidity might be called into question: Rigid habits might, without individually undergoing any reconstruction, be integrated with highly flexible habits, and these complexes might then produce flexibly adaptive responses, which would hardly amount to any diminution of plasticity. Only a trend toward complexes composed exclusively of routine habits, including those habits that organized the other component habits, would salvage this option for clarifying Kuo's diminution thesis. This trend might be described as moving in the direction of increasing *clusteral rigidity*.

Kuo's own remarks on the subject do not, however, support the idea that his diminution thesis is to be understood in the preceding terms. He says: "At the beginning of development, the organism possesses an indeterminate, wide, and extensive range of potential behavior patterns, only a small number of which are actualized during the process of development. All the other potentials are gradually, one after another, prevented from realization due to the gradual reduction of plasticity, which eventually reaches the critical point."[41] Ignoring the hint of circularity, the possibility that a reduction of plasticity might be defined by Kuo as the potentials' being prevented from realization, we might well inquire what Kuo means by a "critical point." He wants the term used in something like a physicist's sense:

> that is, the temperature above which a substance in gaseous form cannot be liquefied no matter how much pressure is applied. Similarly, in the developmental process certain patterns of behavior may have been fixated to such a degree that no effort to reorganize or repatternize, or even modify them, can be successful. [Thus, in cases such as that of] a gray quail . . . raised in isolation for a period of time beyond which it could no longer be socialized . . . we would say that the animals . . . have passed the *critical point . . . of elasticity* or *flexibility* for behavioral reorganization."[42]

The sorts of cases, the fixated behavior patterns, Kuo has in mind are very general or broad; and though they might be said to exhibit some sort of rigidity, they might also be thought to show notable plasticity. Thus, asocial behavior— of the sort exhibited by the gray quail Kuo mentions—might at once be resistant to efforts at eliminating it and take many different forms in diverse social circumstances. And given this sort of plasticity, it becomes questionable whether one could reasonably construe the coexisting sort of rigidity as *clusteral,* as resulting from complexes of purely routine habits.

The fixated behavior patterns Kuo takes to be responsible for the diminution effect might well be described as general character traits or, in the absence of anything resembling a character, general behavioral traits. And though it would be wrong to accuse Kuo of supposing that anything an organism does diminishes its potential, by virtue of precluding its doing something else, there is much truth to the idea that he takes many general behavioral traits to preclude, and thus diminish, the organism's potential for contrary traits.

4.8 Social and Structural-Functional Influences on Plasticity

Kuo's diminution thesis is perhaps best understood not as the positing of a nomological developmental trend, either absolute or conditional, but as a far more tentative generalization to the effect that "development tends to actualize certain potential patterns to the exclusion of others." [43] This generalization is based upon some of Kuo's own empirical studies. But even as they indicate a tendency for plasticity to diminish, these studies attest to the enormous plasticity of behavioral patterns that might have been otherwise fixated, in any of a number of very different, sometimes opposing ways.

To avoid some likely confusion about what Kuo has to say, it may be helpful to distinguish more explicitly than he does between the plasticity of behavior generally and the plasticity of specific varieties of behavior. In the terms of Kuo's theory, the former sort might be further identified as the *plasticity of overall patterning,* a plasticity that is initially enormous but inclined to diminish; and the latter sort might be identified as the *plasticity of partial patterns* (or of behavioral traits). The plasticity of overall patterning is its capacity to pick up partial patterns; their plasticity, in turn, is their "flexibility" qua specific behaviors within a repertoire.

To sum up some of his major findings about what he calls social plasticity (or the plasticity of patterns of social behavior), Kuo says: "The wider the range and variety of the environmental context, the wider the range of the behavioral repertoire and the more flexible the behavior patterns." [44] Now although this generalization might seem to concern only the plasticity of partial patterns (of social behavior), Kuo's more detailed statements of those effects upon which the generalization is based suggest that it is also meant to apply to the plasticity of overall patterning. Here are Kuo's statements of four effects of rearing "in an extremely narrow environmental context" [45*] and (selected) illustrations of each: "(1) total or partial loss of plasticity in reorganizing behavioral patterns to meet the demand of a new environmental context." Thus: "The early establishment and fixation of a monotonous diet for turtles . . . , birds, cats, and dogs . . . make it difficult for the animal to change its eating patterns; the longer the fixation the greater the difficulty in changing." "(2) narrower range of behavioral repertoire." Thus: "The range of song repertoire is greatly influenced by the range of social contacts with other birds, even merely by hearing them in a Chinese tea house." "(3) an easy disintegration and difficult recovery of originally acquired pattern when confronted with some variation of the environmental context." Here he gives the illustrative example of "the abolition of long-established fighting habits in birds by merely introducing a small object or sound; when such birds were placed in a bird colony, they displayed panic behavior for many hours. These birds had great difficulty in redeveloping social bonds, although kept in the colony for many months." "(4) development and fixation of abnormal and antisocial behavior patterns." Thus: "Fighting in fish, birds, and dogs, asexuality in birds and dogs, hostility toward strangers, etc., are generated when the animals are brought up in a controlled environmental context with very little variety of social contacts. Training for fighting, asexuality, and other antisocial behavior would not succeed if the animal grew up in a varied environment." [46]

As if to confirm the importance of distinguishing between the plasticity of overall patterning and that of partial patterns, this last finding suggests that the former plasticity may even be decreased in consequence of an increase of the latter; for if sociability may be viewed as a highly plastic (or flexible) partial pattern, then it may be seen here to prevent the realization, via training, of a host of other, far more rigid partial patterns (of fighting, etc.). But this conclusion from the last finding might make one think twice before suggesting, as Kuo seems to do, that narrow environmental contexts are especially likely to promote the diminution of plasticity (of overall patterning). A flexible response (e.g., an omnivorous eating pattern) may be as prone to fixation as a rigid one (e.g., a mackerel-with-rice habit) and as likely to preclude the other pattern: Thus, a cat with the flexible response might be unable to acquire the rigid one, unable to come to prefer starvation to a soybean diet.

Kuo's diminution thesis, properly reconceived, would seem to amount to the idea that you cannot teach an old dog new tricks at odds with its old tricks, re-gardless of how flexible those old tricks might be. This is not, shades of the tree image of potentials, some misconceived notion that actuality precludes potenti-ality. And, despite Kuo's apparent equation of reduced plasticity with potentials' being prevented from realization, his notion of fixation would seem to save the thesis from circularity—assuming that this notion gets cashed out in a substantive theoretical way. Pressed on the point of whether fixation is irreversible, whether critical points are truly points of no return, Kuo would seem quite content to countenance any reliable findings to the contrary. For rather than serving as the cornerstone of his theory of epigenesis, the diminution thesis seems to function primarily as a way of accounting for certain rigidities of behavior in terms other than those of species specificity and innateness; and evidence contrary to that for fixation might only further undermine nativistic ethology.

4.9 Kuo's Program of Charting Ranges of Behavioral Potentials

Kuo contends that a major task for the epigenetic behaviorist is to "accept the morphological structure of an organism at a particular stage of development as given and proceed to work out experimentally the potential range of behavior patterns." [47*] This is to be done by attempting to create "behavioral neopheno-types"—patterns of behavior that are "novel in the sense that the possibility of their existence has not been recognized by the naturalists or other students of animal behavior." [48] Each of these patterns will be one more potential of the organism.

Gottlieb describes this research program as Kuo's way of trying to gain "knowledge of the outside limits of the range of behavioral potentials." [49] This construal has apparent advantages: Were the range arrayed along a line, a discovery of the range's end points would determine its extent. But on the abstract face of it that range is treelike (rather than linear), and each behavioral neophenotype is but one path along that tree. Accordingly, talk of outside limits loses its boundary-setting force: We could not hope to gauge the overall shape or extent of the abstract tree by discovering its left-most and right-most paths, since the relative locations of paths on such a tree would be highly arbitrary.

The task of attempting instead to discover all the paths might seem more sensible, albeit unmanageably large. Such a task obviously could not be undertaken for a single organism, since each individual (however large or small its tree) gets only one path, one behavioral life to live; but Kuo might well set about producing as many different neo-phenotypes as possible within some highly inbred groups of animals. Would this afford a reliable indication of the range of behavioral potentials of any individual morphological phenotype? Only if inbred animals may be presumed to share the morphological features that are pertinent to the ontogenesis of behaviors; so let us consider the merits of this presumption, by delving further into epigenetic views about the relationship between morphology and behavior.

Kuo's longest-standing opposition to instincts is based on some findings about the interrelated influences of morphological structures and their functioning upon the ontogenesis of behavior. His own studies of chick embryos have convinced him

> that the shape, size, and proportional length of head, neck, limbs, and trunk, especially in the early stages, determined the patterns of movement in the avian embryo. . . . that the heartbeat in the embryonic chick is initiated in the part of the primitive heart in which morphological differentiation is most advanced and the metabolic rate and glycogen concentration highest. . . . that ontogenesis of behavior, in the avian embryo at least, is initiated by the heartbeat and that the origin of rhythmic movements of the embryo during incubation may be traceable to the rhythms of cardiac movements.[50]

Naturally enough, Lorenz insists that unless a further appeal is made to genetically determined neurological structures, Kuo cannot hope to explain on the basis of such findings "why only certain birds peck after hatching, while others gape like passerines, dabble like ducks, or shove their bills into the corner of the mouth of the parents, as pigeons do, although they all, when embryos, had their heads moved up and down by the heartbeat in exactly the same fashion." [51] But Kuo rejects the ethological designation of these distinctive behaviors as "species-specific"—a term that, as used by Lorenz, would suggest that they had resulted from genetically determined neural organizations. Kuo prefers to call such behaviors "species-typical," and he maintains that "other things being equal, the general body framework and the detailed structures of the oral and vocal apparatuses and the extremities determine the modes of locomotion, vocalization, eating, drinking, etc. of the species." [52] In still more direct response to Lorenz, it may also be observed that the difference between dabbling and pecking is chiefly a matter of what is done *where* with a duck's bill as opposed to a chick's beak.

Nevertheless, Kuo's account might seem to concede that the behaviors in question are, even if not in every whit instinctive, genetically predetermined—given that gross morphology is. But what if some doubt could be cast on the idea of a *normal* embryonic environment in which morphology unfolds? Some fascinating food for skeptical thought is supplied by D. S. Lehrman: "If embryos of the fish *Fundulus heteroclitus* are kept in magnesium chloride solutions, a small percentage of them will develop into hatchlings with only one centrally-located eye . . . , and these fish will apparently be able to see." [53] Before taking the line Lorenz would, dismissing the case as one of purely pathological development and, so, of little interest to students of evolutionary adaptation, we should consider some questions Lehrman poses: "First, are the number and location of the eyes an adaptive character; and second, is the information about the number and location of the eyes located in the genome or in the relationship between the genome and the chemical environment?" [54] Indeed, apropos of the first question, are *which* number and location of eyes adaptive? Let us suppose that in "normal" posthatch environments two eyes are more adaptive than one, but that in "abnormal" posthatch environments having high concentrations of magnesium chloride, the one-eyed fish is king. Perhaps fewer foes survive the salt-polluted waters, so two-sided lookout is not as important, while the one eye is keener and/or less subject to the salt's irritating effects. Who knows what might prove adaptive in ever more polluted environments of the future? The king might quickly lose his crown, however, should he be transported, after hatching, to what was once deemed the "normal" environment for the species to which our apparently royal fish can still claim membership. One moral of this story is that "normal" and "abnormal environment" are not fixed concepts.

Of course despite the esoteric possibilities just mentioned, there does seem to be some undeniable merit to the idea of a normal range of embryonic environments. And this normalcy could well be said to be secured biologically, sometimes at the expense of a dramatic depletion of bodily reserves for the environment-providing female. This normal range of environments suffices to ensure gross morphological similarities among members of the same species and, so, would seem to support the idea that even "species-typical" behaviors are largely innate. Nonetheless, that same range might not ensure the development of fine-grained, neuromorphological similarities; for even very slight variations within that range might effect differently detailed neurological organizations among genotypically similar embryos, even clones.

In turn, the differential effects of even slightly different neural structures on behavior might be significant. Thus, even if the general patterns of species-typical behaviors were the same throughout the whole range of neurological variation, important differences of behavioral detail (such as speed of locomotion, frequency of eating, amount of drinking, etc.) could remain. And when it comes to other, less typical behaviors of a given species, the effects on behavior might be far more profound. Accordingly, it might be said, with reference to the issue motivating the present discussion, that perhaps Kuo should give up on the whole idea of attempting to chart ranges of behavioral potentials; for those ranges could vary widely if neurological organization were not constant. Only of an individual organism might one reasonably "accept the [detailed neuro]morphological structure at a particular stage of development as given"; but only of a large group of organisms would it make partial sense to suppose that one might "work out experimentally [a] potential range of behavior patterns."

4.10 Rethinking the Role of the Brain in Behavior

The above argument against Kuo's range-charting program stresses the role of neuromorphological structures in determining an organism's range of behavioral potentials, but Kuo sometimes seems to downplay that role, even to reverse the received order of explanation:

> If future experimental investigations show that the structural-functional capabilities of the human brain are far more complex than those of the brains of nonhuman primates, it is because the potential and actual range of man's behavioral gradient patterns is broader and more complex. The epigeneticist may be justified [in] tentatively assuming that the C.N.S. acts merely as the excitatory, inhibitory, and coordinating center for the activities of other parts of the body in the whole gradient system of behavior.[55]

Kuo does not mean to deny that neurological structures determine behaviors so much as to assert that other factors are at least as important as those structures, to emphasize that the entire epigenetic system does the determining. What Kuo means to deny is that typically Western, "conventional conceptualization" according to which the brain is "the seat of mind, intelligence, innate behavior, memory, learning, motivation, emotion, etc."[56] Indeed, he speculates that "if we could succeed in exchanging brains between a human neonate and a gorilla neonate and raise them in an identical environment with complete absence of human language and human culture, the human child would grow up to behave with human characteristics and the gorilla with the characteristics of its own species because the skeletal framework of the body in general and the fine structures of the hands of the two species are different."[57]

Kuo thus proposes that even in the absence of (allegedly) predetermined, species-specific neurological structures, the larger epigenetic system (and, especially, human anatomical structures) would afford the potential for some typically human behaviors. Notice that no apparent appeal is here made to the bidirectionality thesis and that Kuo may tacitly concede that a human with a gorilla's brain could not aspire to language and culture. Kuo alludes to the idea of form's following function only in connection with his converse speculation: "If an animal were given a human brain, but not the framework of the human body and other particularly human anatomical structures, it would probably not develop typically human behavior; moreover, at least part of the brain tissue would probably be undifferentiated and unfunctional, or even become atrophied, because, without those complex human activities that are possible only with the mechanisms of the human form, there would be no effective stimulation." [58]

Although one could hardly fault Kuo with a failure of nerve, given these astounding speculations, it might be well to inquire about why he seems not to countenance the still more radical possibility that a human with a gorilla's brain might, afforded with the stimulation of complex human activities, end up with more highly differentiated and functional brain tissues than any gorilla, even one with a human brain. The reason, I suspect, is that Kuo is too thoroughly committed to the notion of genetically preestablished ranges of behavioral potentials. This commitment leads him to accept an overly restrictive version of the bidirectionality thesis, a version according to which form follows function only *as far as heredity allows*. Thus, the passage just cited seems to view a lack of stimulation as leading to a lack of full neurological development, as if a particular neuromorphological pinnacle were genetically predetermined.

Kuo should dispense with the idea of a predetermined range of behavioral potentials, if only because that idea does not square with his own allegiance to methodological behaviorism: Controlled observation and experiment could not discover such a range. In terms of Kuo's own analogy of a log's potential, one might pointedly observe that setting about to make as many different things as possible with nearly identical logs would never serve to chart a determinate range of any such log's potentials. There may be some plausibility to Kuo's comparison between his attempts to create novel behavior patterns and the synthesizing efforts of hydrocarbon chemists, but Kuo himself notes that the possibilities of new forms in both realms "appear to be almost unlimited";[59] so perhaps there is no *determinate* range of behavioral potentials to chart.

There may also be some merit to Kuo's claim that his own efforts to create behavioral neophenotypes have demonstrated that "the potential range of the repertoire of behavior in each individual is far greater than assumed." [60] But Kuo is wrong to put so fine a point on the concept of a *potential* range of behavior. It is, after all, his own contention that the *actual* range of an individual's behavior is not preestablished: "No individual animal, especially among the higher vertebrates, is born with a genetically predetermined range of behavior patterns aside from the anatomical and functional limitations of certain specific organs." [61] Why, then, should he be any more sanguine about the idea of a genetically predetermined *potential* range of behaviors?

Were Kuo to give up on this idea, he might also begin to see some merit in denying (or not accepting a priori) the related idea that behavioral functioning is limited in its morphogenetic role to helping or hindering the attainment of an organism's full neuromorphological potential. The result of not subscribing to this unwarranted restriction on the bidirectionality thesis would be a still more extreme theory of epigenesis, a theory that allowed for ever wider, unpredetermined ranges of potential behaviors. Such a theory does not deny the importance of genetic factors in morphogenesis, but it does suggest that fortuitous happenstance might play a more significant role in the ontogenesis of behavior than talk of a preestablished range of behavioral potentials would seem to suggest. What seems at most a minor departure from Kuo's theory is a major move away from any concession to instinct.

4.11 Jensen's Research Program: Comparative Harmony, Ethological Discord

One research program that accords well with an epigenetic outlook was once proposed by Donald Jensen. Here is Lehrman's convenient summary:

> Jensen had suggested that many operations other than genetic selection or training could produce differences in behavior between animals. These operations include nutritional variations, alterations of the nervous system, hormone treatments, etc. Jensen suggested that studying the effects of a wide range of differential treatments upon the development of behavior differences, with the intention of inductively integrating the information thus acquired, would be a more fruitful and less controversial way of coping with problems of ontogeny and of causality than the prevalent attitude of treating the question "innate or learned?" as a primary and ultimate question on which all others must hinge. [62]

Jensen's belief that his research scheme would be not very controversial was overly optimistic: Lorenz testily dismisses it as "a Herculean task promising only uninteresting results." [63] Lorenz, committed as he is to his own plan for discovering the organism's preestablished adaptedness, insists that "[as] students of behavior, we are not interested in ascertaining at random the innumerable factors that might lead to minute, just bearable differences of behavior bordering on the pathological." [64] He counterproposes that "all we have to do is to rear an animal, as perfectly as we can, under circumstances that withhold the particular information which we want to investigate. We need not bother about the innumerable factors which may cause 'differences' in behavior as long as we are quite sure that they cannot possibly relay to the organism that particular information which we wish to investigate." [65]

Lehrman characterizes this dispute as one between scientists interested in causal mechanisms of ontogenic development and scientists interested in adaption and evolution. He also extends an olive branch to ethologists: "The student of adaptation and evolution may . . . be talking in entirely legitimate and meaningful terms about the problem in which he is interested." [66] However, there may be reasons to question the legitimacy of some of Lorenz's terms (e.g., "information"), and it may be that the conceptual overlap between the two sorts of interests is such that it makes no sense to attempt to pursue Lorenz's interests independently of ones like Jensen's. Lorenz would have all students of behavior first discover each species' inventory of innate behaviors and only then worry about discovering the causal mechanisms that give rise to them. But he seems unprepared to concede that Jensen's approach would be a suitable way of investigating such mechanisms, for Lorenz's hasty assumption that pathology is interesting only to pathologists seems to discount the possibility that the study of pathology may help scientists to discover the determinants of nonpathological developments.

Of course even if Lorenz did accept Jensen's approach as a way to investigate mechanisms underlying *antecedently identified* innate behaviors, one might still question the legitimacy of the apparently sharp division of scientific labors thereby suggested. Although Lorenz acknowledges the right of other scientific inquiries to go beyond ethological findings, he doesn't seem to allow for the possibility that such inquiries might discredit those findings. Lorenz views the ethologist's inventory of species-specific behaviors not merely as a tentative guide for further research, but as bedrock, the indisputable foundation for future investigation.

The ethological inventory, without benefit of any investigation of the actual mechanisms underlying certain behaviors, presupposes that those mechanisms are devices for the implementation of genetic information about the environment and how to behave within it. But would such a theory-laden inventory afford even a provisionally useful guide for research into causal mechanisms of behavior? About all that might be said in the inventory's favor is that it affords some prior identification of behaviors for which mechanisms might be sought. But the fact that those behaviors are further classified as innate should be of little consequence to the conduct of that search. That classification would not, for example, justify dispensing with the experimental manipulation of the "normal" environment, on the grounds that it is just this environment that serves to construct the behavior-producing mechanisms according to their genetic blueprints. To the contrary: Such experimentation might be the very means of discovering how that construction actually takes place. And, of course, since we may hardly be said to know in advance of that discovery that the construction is purely maturational, such experimentation might also be the means of deciding whether or not the behaviors really are as innate as the inventory presumes them to be.

Jensen's program neither needs nor benefits from Lorenzian inventories—it renders them superfluous. If we consider that program from an epigenetic point of view, such inventories seem all the more pointless: So, for example, our very sense of what constitutes a given piece of behavior, a partial pattern of gradients, might not survive an inquiry that has the potential for discovering divers mechanisms underlying behaviors that are superficially (i.e., at the level of explicit gradients) alike. And, more generally, it might be observed that while positive findings of the deprivation experiment do not suffice to disprove Kuo's epigenetic theory, the theory, if confirmed by Jensen's program, would suffice to discredit the findings. Those findings simply do not preclude the possibility that the (presumptively innate) patterns are, in truth, products of a far more complex and variable determination "by the total environmental context, the status of anatomical structures and their functional capacities, the physiological (biochemical and biophysical) conditions and the developmental history up to that stage." [67]

Suppose that, by subtle environmental modification, it is possible to produce behavioral neophenotypes some of whose response patterns are contrary to those identified (after numerous applications of Lorenz's deprivation technique) as species specific. Before dismissing these contrary patterns as pathological, as Lorenz might be wont to do, we ought to consider how viable the organism is, how adapted these patterns are. If the organism with these deviant patterns proves well adapted to the "abnormal" environment that effects them, then, whether or not those patterns are accompanied by other gross (morphological, etc.) departures from the norm, it is entirely possible that the supposedly abnormal environment was the very sort that, during phylogeny, selectively favored that organism's genotype.

Even used merely as tentative guides for research, Lorenzian inventories are potentially misleading. A researcher relying on such a guide would attempt to work out the details of underlying mechanisms for the behaviors provisionally labeled as innate. Such a researcher might notice that other behaviors result from exposing the developing organism to modified environments, but these other behaviors, which might be adaptive in their own environments, could well be dismissed as uninteresting anomalies, results of pathological disturbances of normal development. By casting the researcher's task as one of searching for mechanisms underlying innate behaviors, the tentative guide may set up certain expectations about what sort of findings are worth bothering about and thus prejudice the inquiry in favor of innateness hypotheses.

Jensen's program seeks to avoid such theoretical tendentiousness, to circumvent any prior questions about the innateness of behavior. And although it harmonizes with an epigenetic outlook, this program cannot be charged with presupposing an anti-instinctivity stance, either. The program is not even as tendentious as Kuo's

program of creating behavioral neophenotypes, which swears prior allegiance to
the idea of a preestablished range of potentials. Jensen's approach aims instead to
produce all sorts of behaviors, new and old, in hopes of discovering their causal
antecedents, whatever they may be. The fault of this program lies, if anywhere,
in its overly restrictive methodology: By proposing merely to integrate findings
(about the effects of differential treatments on behavioral developments) inductively, Jensen would seem to ignore the value of more hypothetico-deductive
strategies of research.

4.12 Chief Epigenetic Challenges to the Ethological Enterprise

Kuo's epigenetic view is perhaps less a fully articulated theory than a tentative
scheme for organizing findings (such as those of a Jensenian program) into a
coherent conceptual framework. Although his scheme is too loosely hypothetical to allow for the deduction of testable predictions, Kuo can be said to have
advanced, methodologically, beyond mere inductivism. And although his own
researches have concentrated on the arguably misconceived program of charting
ranges of behavioral potentials, Kuo does urge the prospective epigeneticist to
engage in a wide variety of other, interdisciplinary researches aimed at obtaining
"a comprehensive picture of the behavioral repertoire of the individual and [that
repertoire's] causal factors from stage to stage during development." [68]

The results of Kuo's own empirical efforts may be sketchy, but they make
for a fairly strong case against the pro-instinctive stance of some major ethologists. Apart from its vain purpose of gauging the ungaugeable, Kuo's program
of creating behavioral neophenotypes has led him to conclude, on a rather firm
empirical basis,

> that it is equally "natural" for a predatory bird, an eater of small birds in
> the wild, to become very fond of them in confinement; that a cat is born
> not only a rat-killer but a "rat-lover" as well; that a dog is a born bird-lover
> as well as a born bird-hunter; that mynahs, quite unconfined, may prefer
> hopping around on the ground competing for food with chickens to flying
> out to search for food . . . ; that a dog can be a good swimmer and "enjoy"
> a shower but be afraid of a little rain; that a sexless dog could be developed
> without altering the dog's original nature.[69]

Such a list is likely to confound many an inventory of species-specific behaviors, of instincts.

As a competitor of Lorenz's ethological outlook, Kuo's epigenetic view emphasizes almost every other determinant of behavior besides heredity. Kuo goes
so far as to suggest that the evolution of behavior need not be understood in

Darwinian-genetic terms. He claims that behavior evolves within a species whose genotypes (or "gene pools") remain the same. This evolution is said to be especially evident in higher social animals, "whose patterns of social behavior are shaped mainly, if not wholly, by the developmental history in conjunction with the ever-changing environmental context." [70] But even in animals on "lower phyletic levels . . . such as arthropods, particularly the social insects, shifting from one ecological niche to another may induce changes in certain behavior patterns. As long as the general nature of the new environmental context remains relatively unchanged, despite inevitable variations, we may expect that the newly induced behavior patterns, or behavioral neo-phenotypes of the group as a whole would be carried on from generation to generation." [71]

Phylogeny, on Kuo's view, is only indirectly relevant to the ontogenesis of behavior. Phylogeny does set morphological limits on the kinds of behavior that might be developed, though as Kuo's efforts to produce neophenotypes tend to suggest: "The actual range of the behavioral repertoire is determined by the history of the animal's development in relation to the environmental context. Because of its anatomy, the Peking duck has the potential to swim, but whether it will actually become a water bird or develop a water phobia depends on the environmental context under which the newly hatched duck is reared." [72]

Part Two

Philosophical Remonstrances

Man the machine—man the impersonal engine. Whatsoever a man is, is due to his *make,* and to the *influences* brought to bear on it by his heredities, his habitat, his associations. He is moved, directed, COMMANDED, by exterior influences— *solely.* He *originates* nothing, not even a thought.

—*Mark Twain*

Chapter 5

Plasticity, Teleology, and Instinct

Although the discussion to this point could lay no claim to having disproved the existence of instincts, some serious conceptual and methodological weaknesses in instinct-oriented theories and research have been exposed. On a less negative note, some promising explanatory alternatives to pro-instinctive theorizing have been found. Still, the clash with instinct is incomplete.

In its efforts to secure full opposition, this discussion will move in an ever more philosophical direction. Destructive purposes will here be served by a constructive project, the development of a position more thoroughly antagonistic to instinctivity than any thus far considered. Such construction is often assumed to be the prerogative of science, not philosophy, but perhaps that assumption bears further examination. In any event, what follows is not just a skeptical critique of instinct, but also a positive philosophy of action and of mind.

5.1 S-R Theoretic Plasticity

A recurrent theme of Part One was that the plasticity of behavior is, variously, at odds with its instinctiveness. To gauge the depth of this apparent opposition, let us examine what might appositely be meant by *plasticity*. Kuo's epigenetic view, which was given the last word in Part One against instinct, purports to be close in spirit to stimulus-response psychology. So it might be instructive to consider, first, how plasticity might be understood in ordinary S-R theoretic terms and, then, whether Kuo's epigenetic view is consonant with such an understanding.

If behaviors are construed as responses to stimuli, then the plasticity of a given behavior might be said to consist in its being a response to more than one (type of) stimulus. And the broader/narrower the range of stimuli capable of eliciting a given response, the more/less plastic that behavior. Such plasticity, call it "many-one plasticity," might be a valuable addition to many S-R theories, including a chained-reflex account of instinct. Koffka, as noted earlier, argued against that account on the grounds that there would be an endless number of distinct possibilities of stimulation from even a single stimulus object (e.g., a buzzing bee to which a spider responds by fleeing). He maintained, in effect, that this number of stimulations would demand an equally and implausibly large number of *nonplastic,* predetermined S-R links. Koffka's alternative was a Gestalt-theoretic departure from the S-R viewpoint. A less drastic departure might be to attribute many-one plasticity to the S-R links.

But attractive though this option appears, it might be difficult to reconcile this notion of plasticity with a commonplace S-R theoretic definition of *same stimulus* as "stimulus capable of eliciting the same response." Such a definition precludes the possibility that a *range* of stimuli might elicit a given response. To get around what may seem to be a purely verbal hurdle, let us consider other ways of putting the point. Suppose it were suggested that the notion of a stimulus is allomorphic— that any of a class of functionally identical but observationally distinguishable items constitutes one stimulus. Problems abound: Distinguishable by whom, by what means? Surely by the perceptually unaided responding organism; for if, say, slight variations in the frequency of sound waves were observable only with the aid of technologically advanced instruments, those variations might not exist for the organism that responds identically to them all. But, then, a further problem arises when we try to get at what sorts of differences among putative allomorphs do exist for the organism: Evidence for distinguishability would be drawn from differential responses to numerically distinct items, and that would establish distinguishability only on pain of also indicating a lack of functional identity. Of course one might try to get at distinguishability by studying a variety of response types besides the one type for which the items are functionally identical. Thus, a bird might peck indiscriminately at red and orange seeds yet respond differentially to red as opposed to orange eggs. But this might merely indicate that although in some contexts the two colors can be distinguished, in other contexts no differences can be observed. And even if the bird could be trained to respond differentially to red and orange seeds (e.g., to peck at one but not the other), it might be suggested that this fails to get at whether the colors were originally distinguishable or are, instead, distinguishable only after training. There is, it seems, some indeterminacy of observation, and, in consequence, this notion of plasticity is not easily squared with a philosophy of methodological behaviorism.

But the prospects for a reconciliation might improve if we knew the workings of the proverbial black box, the causal mechanisms connecting stimuli to responses. Suppose we were to discover a range of different causal chains, each of which begins right after a different stimulus and all of which end just before the same response. Such a response could well be said to exhibit many-one plasticity. However, if different stimuli were followed by some *one* chain as well as some *one* response, then no such plasticity would be manifest. (This could happen, say, if a bull becomes identically enraged whenever optically exposed to light waves within a given range of "red" frequencies, and it turns out that the bull's optic nerves respond identically to all the frequencies within that range.)

Another sort of plasticity that might be expressed in S-R theoretic terms will be called "one-many plasticity." The idea is simply that one stimulus can give rise to more than one response. And the wider the range of responses to a given stimulus, the greater this plasticity. Plausible cases of such plasticity are easy to come by:

Even a simple knee-jerk response will, at a microlevel (which considers, e.g., the individual muscle fibers that contract), doubtless be slightly different each time it occurs, however similar the eliciting stimuli might be. And at a macrolevel, the phenomenon of habituation, wherein repeated exposure to a given stimulus causes the organism's responsiveness to wane, is a clear case of multiple responses to the same stimulus. Any sound S-R theory would do well to countenance such plasticity, and it should be noted that doing so would not be a departure from a straightforwardly deterministic view of behavior. The hypothesis of one-many plasticity does not require that behavioral responses fail to be uniquely determined by their causal antecedents, for these antecedents include the inconstant organism as well as the unvarying stimuli that prompt it to react as it does.

5.2 Epigenetic Plasticity

Although Kuo's epigenetic viewpoint is supposed to be thoroughly behavioristic, it departs from S-R orthodoxy by extending both the notion of behavioral *responses* (to encompass all variations, all "behavioral gradients," macro and micro, of the organism) and the notion of *stimuli* (to include a larger "environmental context"): "We shall define environmental context as a physical and social (in the case of social animals) complex surrounding the animal, of which the effective stimulus or stimulating object is a component or integral part. Like behavioral gradients, the environmental context is a variable and complex dynamic system organized into definite patterns. Except under very strict laboratory control . . . , the environmental context is never static." [1]

Yet although Kuo stresses the dynamic interplay between these two complementary "variable and complex dynamic system[s]"—that is, the *environmental complex* and the (system of) *behavioral gradients*—he does not favor a Gestaltist (or, field-theoretic) approach to the study of this interplay. To the contrary, Kuo insists that his very concepts of these interacting systems "are not only experimentally analyzable, but can be analyzed only by the atomistic methods that have been anathematized by the Gestalt psychologists." [2] Accepting this promissory note for an atomistic analysis, we might also expect to gain along with it an S-R theoretic understanding of plasticity. And, indeed, the above notions of plasticity already seem to fit in quite well with both S-R and epigenetic theorizing. Thus, the causal chains that helped to explicate the notion of many-one plasticity may be viewed as sequences of implicit behavioral gradients, thereby bringing out more clearly the S-R theoretic character of that notion as well as showing its epigenetic stripes. And much the same can be said for the notion of one-many plasticity, since the intraorganic changes that may be presumed to account for the multiplicity of responses to a single stimulus are also interpretable, in epigenetic terms, as behavioral gradients.

Still, a couple of features of Kuo's theorizing might not seem consonant with the above two notions of plasticity. First, Kuo's contention that an ever-changing environmental context is the closest epigenetic approximation to what is called a stimulus makes it difficult to regard the notion of one-many plasticity as compatible with his theory. But since Kuo does allow that the environment might be rendered static "under very strict laboratory control," this problem could be dismissed as one arising only in (almost inevitably) loose practice, not in principle. A second feature of Kuo's theory that seems to pose related difficulties is his bidirectionality thesis:

> In ontogenesis, both patterns of behavior and patterns of the environment affect each other and are therefore in a constant state of flux; that is, changes in the environmental patterns produce changes in behavior patterns which in turn modify the patterns of environment. The epigenetic view of behavior is bidirectionalistic . . . rather than environmentalistic, as it considers every behavior pattern as a *functional product* of the dynamic relationship between the organism and its environment, rather than as a passive result of environmental stimulation.[3*]

So, while both many-one and one-many plasticity are fully reactive, deterministic notions, quite consistent with an S-R theoretic understanding of behaviors generally, Kuo's bidirectionality thesis might seem to suggest a rather more proactive, nonpassive understanding of behavior and its plasticity. But a closer look at what Kuo is saying reveals no abandonment of deterministic ideals: Although he does deny that environmental stimulation alone produces specific behavior patterns, he affirms that they are functional products of—and so, presumably, uniquely determined by—the values of *all* the component variables of the epigenetic system. And the "constant state of flux" that Kuo predicates of all patterns within the epigenetic system, encompassing the organism and its environment, is for him a flux that can in principle be analyzed into those component variables. Once again, no conceptual incompatibility between Kuo's theory and the (above explicated) two notions of plasticity has been discovered.

Nothing seems to preclude the construal of epigenetic plasticity as an assortment of the (essentially reactive) many-one and one-many varieties of plasticity. To encourage the acceptance of this construal, I should now like to relate it to an exegetical point made earlier (see section 4.8): In trying to establish a reasonable connection between epigenetic behavioral patterning and plasticity, I suggested a need to distinguish between the *plasticity of partial patterns* (or behavioral traits) and the *plasticity of the overall patterning* of the epigenetic system. I should now like to suggest that the plasticity of partial patterns (e.g., an eating response elicited by many different sorts of food) may be interpreted as a many-one plasticity, while the plasticity of the overall patterning (e.g., a cat which is as much a born rat-lover as a rat-killer) may be interpreted as a one-many plasticity. The difference, simply put, is that between the plasticity of a single response type capable of being elicited by more than one stimulus type and the plasticity of an organism capable of having more than one response type to a single stimulus type.

5.3 Teleological Plasticity

Since instincts are, by repute, goal directed rather than merely reflexive behaviors, the preceding, S-R theoretic understanding of behavioral plasticity might seem irrelevant to whether plasticity is at odds with instinctivity. To join this issue, one needs a more teleological sense of plasticity. Such an understanding might be afforded by R. B. Braithwaite's account of teleology *as* plasticity.[4*]

Braithwaite tries to analyze only those teleological behaviors that are *not* consciously goal seeking, those behaviors whose "goal directedness consists simply in the fact that the causal chain in the organism goes in the direction of the goal."[5] The plasticity of an organism (or other "system") with respect to its goal is, roughly, a matter of the organism's being able to attain that goal under varying conditions, often via different causal sequences. The detailed analysis is this: Suppose that each causal chain of events, c, in a system, b, is uniquely determined by its initial state, e, and a set of field conditions, f. *Gamma* is the set of all causal chains (in b) that have a g-goal-attaining property—that is, the property both of ending in an event of type g and of containing no other event of that type. And "the variancy [*phi*] is . . . the class of the f's which uniquely determine those c's which are members of [*gamma*]."[6] Finally, a system, b, has *plasticity* (with respect to a goal, g) whenever *phi* is greater than one—which is to say, whenever that system (in state e) could produce g-goal-attaining causal chains in varying circumstances, under any of two or more sets of field conditions. And the larger the number of sets of field conditions under which a system could produce such chains, the greater its plasticity.

Braithwaitean (or teleological) plasticity is comparable to many-one plasticity: In S-R-theoretic terms, the field conditions are stimuli, and the goal is a specific behavioral response; in epigenetic terms, the field conditions are environmental contexts and the goal is a single partial pattern of behavioral gradients. But there is a notable difference: According to Braithwaite, "The essential feature . . . about plasticity of behavior is that the goal can be attained under a variety of circumstances, not that it can be attained by a variety of means."[7] With respect to many-one plasticity, though, the variety of the causal chains linking (many) stimuli to some one response was deemed paramount; and plasticity was said not to obtain in cases where divers (sets of) stimuli occasioned identical chain types. The thought behind this denial was that a difference (in stimulus or environmental context) should make a difference to the organism, else its virtue of plasticity might amount to no more than an inability to discriminate. On the basis of this thought, one could also justify modifying Braithwaite's account: True teleological plasticity demands the ability to attain a goal in varied circumstances but also demands a further "variancy" among the g-goal-attaining sequences uniquely determined on the basis of those varied circumstances. And the larger the number of such sequences, the greater the system's plasticity.

Of course even if revised along these suggested lines, teleological plasticity ought to be different from many-one plasticity: The former, unlike the latter, is supposed to involve goals. Yet it is arguable that, on Braithwaite's analysis, any response type involved in a case of many-one plasticity would count as a goal; for to so qualify, it seems, an event's first occurrence must simply be the *terminus* of [varied] causal chains that could be wrought (in system b, state e) by [correspondingly] varied field conditions. Let us consider some difficulties to which this (context-gleaned) conception of a goal is heir.

If the contention that they are termini of causal chains is meant to suggest that goals have no further causal consequences, then goals would not have much utility for teleological systems, either: The attainment of such goals would not even facilitate (via learning, e.g.) subsequent goal-directed behaviors. And if, as Dewey and Kuo maintain, the whole, ever-changing organism is involved in each of its acts, then an organism would have but one dead-end, death-ending, causal chain. Is death, then, the organism's only goal?

An alternative interpretation of Braithwaite is that any single event may be considered a goal, provided only that the event would arise (variously) in consequence of varied sets of field conditions. But, then, the number of Braithwaitean goals would be limited only by our imaginations. Suppose, for example, that a man goes to a tavern to get drunk. Although the man currently resides in a land-locked American city, we might imagine that agents of the British navy will press the drunken man into service against the Spanish Armada. Moreover, we could imagine this to happen to the man in countless circumstances: The man has a knack for going to the very places frequented by H.R.M.'s agents, he is delighted to accept free drinks from strangers, and so on. Therefore, the man has plasticity with respect to being shanghaied; and though in fact conditions are not right for it, whenever the urge to get drunk strikes, this man exhibits teleological behavior directed—unconsciously, to be sure—toward an anachronistic naval career. Such imaginings surely ought to be held in check, yet Braithwaite's scheme, as here interpreted, does nothing to discourage them.

But it may be unfair to take our sense of what Braithwaite supposes a goal to be from his analysis of teleological plasticity. That analysis is used by him as the basis for a further account of teleological explanations, and, presumably, a system should be said to be goal directed, to have a goal, only when such an explanation is correct. Braithwaite associates teleological explanation with prediction. What is predicted is goal attainment (via a causal chain within a system). The prediction is conditional, though not (usually) expressly so: We do not (and often cannot) specify the requisite field conditions. Nonetheless, when we predict that an organism will attain its goal, we are supposing that "the set [*psi*] of field-conditions which will in fact occur will be a member of the variancy [*phi*]" [8]—in other words, that the field conditions that will occur are among those circumstances under which its plasticity enables the organism to—ensures that it will—attain the goal. Accordingly, taking our sense of Braithwaitean goals from the context of such predictions, we are not entitled to credit contemporary dipsomaniacs with the goal of being shanghaied: It is unreasonable to believe that the fantastic circumstances needed to attain that state of affairs is, in Braithwaite's phrase, "at all likely to occur";[9] so it is unreasonable to believe that *psi* is included in (our fancy-generated) *phi*.

But one difficulty, as identified by Israel Scheffler, of thus ascribing goals, on the basis of accurate Braithwaite-type teleological predictions, is that we might then not allow for the possibility of "goal-failure": Scheffler asks rhetorically, "If Fido, trapped in a cave-in, is in fact never reached, is it therefore false that he pawed at the door in order to be let out?" [10] Braithwaite might dismiss this case as one not covered by his analysis of *nonintentional* goal-directed behavior, but the same difficulty could as readily attend even paradigm cases of such behavior: Any extraneous happenstance that prevented a system from completing its preestablished causal sequence would preclude calling the end-that-might-have-been a (Braithwaitean) goal.

Although the difficulty of goal failure does not seem to justify abandoning correct teleological predictions as a basis for ascribing goals to organisms, that difficulty does make it plain that those predictions cannot be the sole basis. But what about Braithwaite-type *explanations*? They might remain a sound basis even when the predictions derived from them are faulty, for a correct teleological explanation might just be one that gives an apt Braithwaite-style description *and* that makes good excuses for bad (or, at any rate, incorrect) predictions based on that description. Thus, we might explain away some goal failures by insisting that the system would have attained its goal, *g,* if only certain factors external to the system at the outset had not prevented *g*'s occurrence. When such an excuse holds true, the system might be described as having been directed at a failed goal.

But what of other cases, ones more akin to Scheffler's example of the dog, in which the organism is doomed from the start to goal failure, given the field conditions then obtaining? Once we begin excusing our teleological predictions on the grounds that the field conditions were not right, even though it would have been unreasonable to suppose that they were right, there might seem to be no end to allowable goal ascriptions based on faulty predictions. But perhaps compelling excuses are available in some such cases—for example, cases in which it can be made out that the causal mechanisms at work in the *non-g*-goal-attaining chain that took place are very closely related to mechanisms operative in other chains that, under more favorable, not unlikely circumstances are *g*-goal attaining. The burden of proof that the mechanisms are closely enough related, whatever that means, rests with the excuse giver, but here is a fairly convincing story: Although the governing mechanism that had to deal with a superheated liquid (mistakenly put in the chamber) did fail to return the temperature within the system to the prescribed level, the operations of this mechanism were quite similar to ones that do, under ordinary circumstances, get the job done; so those operations, though unsuccessful, may still be described as directed toward that goal.

Such excuses begin to exempt Braithwaite-type analyses from the difficulty of goal failure, but another problem, what Scheffler calls "the difficulty of multiple goals" awaits. The new difficulty is supposed to affect a proposal Scheffler volunteers as a way to avoid the goal-failure difficulty. The proposal is to take teleological descriptions nonpredictively, "as attributing plasticity to . . . behavior . . . with respect to the goal indicated." [11] The new difficulty arises when, for example, a cat crouching before a mouse hole would, depending upon the circumstances, as readily avail itself of nearby cream as catch a mouse venturing forth from the hole. Assuming that the cat has plasticity with respect to both goals, though in fact there is no mouse, Scheffler says, "It should be a matter of complete indifference, so far as the present proposal is concerned, whether we describe the cat as crouching before the mouse-hole in order to catch a mouse or as crouching before the mouse-hole in order to get some cream." [12]

Now Scheffler may be right to suggest that his volunteered proposal cannot explain why the former description seems more acceptable than the latter, but what about the above proposal to base ascriptions of goal directedness upon apt teleological explanations (with excuses, when needed)? That proposal would ally itself with the more acceptable of the two descriptions: The crouching would, given the inductive evidence Braithwaite favors to establish that there is some plasticity, be taken to be directed toward mouse catching, since the crouching, presumably, precedes the field conditions (including cream) that subsequently determine cream-attaining behavior. And although the crouching cat may be in a state such that it has plasticity with respect to cream getting (else why should the sight or smell of cream provoke cream-getting behavior?) the crouching per se would not be taken to be that state, only perhaps to be evidence for that state (say, hunger). So the cat's crouching, whether reckoned as its behavior or its initial state, is not plausibly construed, given a Braithwaitean analysis of teleological explanation, as happening in order for cream getting to take place—unless there happens to be evidence that the cat has been taught to crouch, as a trick for which its master rewards it with a saucer of cream.

However, other instances of the difficulty of multiple goals are not so easily dealt with on the basis of Braithwaite's model of teleological explanation. Consider the following example of what might be called a *goal epiphenomenon:* Presented with various kinds of food, an animal chews, variously, and then swallows, noisily. Given Braithwaite's model, the swallowing would (plausibly enough) be construed as a goal, but so (implausibly) would the noise.

5.4 Reasonable Assignments of Teleology

Given the last difficulty, Braithwaite might be accused of failing to capture our preanalytic sense of nonintentional goals. But suppose there is no definite sense of this sort: Say that in speaking of such goals we are extending the use of a term applicable primarily to cases involving human intents and purposes. If so, we might choose to regard Braithwaite's own usage as conferring a special, technical meaning upon that term and to worry otherwise about its philosophic or scientific usefulness. But the situation doesn't quite lend itself to this choice; since Braithwaite does try to capture our prior sense of acceptable teleological explanations, and they would seem to make reference to goals as ordinarily understood. Moreover, even if we do ordinarily understand nonintentional goals by way of extension from primary instances of a more intentional sort, some extensions may be more reasonable than others; so we might still have some standard against which to assess his usage.

Our sense that what I called goal epiphenomena should be distinguished from goals would seem to depend somehow on the whys and wherefores of goal-directed behaviors: Noises caused by swallowing do not have the point to them that swallowing itself does. But how might this intuition be caught by some amplified analysis of teleological explanation? Since Braithwaite's analysis has it that we should rely on inductive evidence for teleological explanations, perhaps a more theoretical approach is called for: the noises would not be significantly efficacious, whereas swallowing is but part of some larger, vitally important digestive process. The idea that genuine goals are thus not causal dead-ends for a system does deal effectively with the case in question, but other cases are not well met: suppose that a machine is devised to address human wants and that this machine achieves its assigned goal but then runs on and on, unstoppably. Events subsequent to the assigned goal would also produce continuing causal sequences, but those events would not seem, intuitively, to qualify as further goals. An event is no less a goal epiphenomenon merely because it determines, within the system to which it belongs, a subsequent chain of events.

Of course one thing lacking in the last case mentioned is a further point to the post-goal-attainment causal sequence. So, let us also make it a requirement of a teleological explanation that the nonintentional goals to which it applies must be causally efficacious in furtherance of their systems' *maintenance, development, or reproduction*. These three features are said, by E. S. Russell, to be "the main biological ends" to which goals of organic activity are "normally related," [13] but nonbiological systems may also have such features. It is a credible suggestion that only of systems having one or more of those features can acceptable teleological explanations be given.

Apropos of the (biological) end of development, consider a case where a genetic mutation occasions a plant's novel tropistic behavior—leaning toward sources of solar *or* atomic radiation. Now suppose that this plant's leaning toward a source of atomic radiation gives rise to an unprecedented, grotesque yet viable growth. Being irradiated would seem, on the basis of the amplified account of teleological explanation, to qualify—and plausibly so—as the plant's goal. But suppose that the plant's genotype endures and other phenotypes with that genotype lack the grotesque development—unless they happen to take root, as their primogenitor did, too close to a radioactive waste dump. Would it still seem plausible to contrue being irradiated as the primogenitor's (developmentally instrumental) goal?

A negative answer might be prompted by the idea that development should be furthered with respect to the phenotypic standard of the genotype: The plant, it might be said, fails in its development to live up to such a standard. Suppose, though, that an enterprising agronomist starts planting all available seeds (of this genotype) next to atomic waste sites—since the grotesque plants are tasty, nutritious, and inexpensive to grow. If phenotypic standards are set with reference to the common course of events, then we would here have to take seriously the idea that being irradiated is a goal of all plants of the genotype, even those that do not have enough nearby radiation to develop "normally."

Of the three biological ends mentioned, maintenance is the most encompassing: Both development and reproduction could be viewed as varieties of maintenance. But this end is also the most nebulous, and goal ascriptions based on (furtherance of) that end could prove equally so. Consider the patently absurd Aristotelean fancy that rocks have a teleological tendency that directs them toward the center of the earth, whenever obstacles to that end are removed. One reason for the apparent absurdity is our post-Galilean sense that mechanistic explanations are more suitable—though if Braithwaite is right, teleological explanations are mechanistic. But another, related reason seems to be the very pointlessness of the posited earthly centerward tendency. Suppose, however, that we construe the earth as a system striving to maintain its sphericity. The rock's movement can then be viewed teleologically, as furthering the maintenance of the larger system to which the rock belongs.

But if this is accepted as a basis for a teleological construal of a system, then perhaps any system whatever may be so construed: Given Köhler's contention that a candle's flame has a dynamic tendency to preserve its shape, it begins to seem that any system might be credited with a measure of self-maintenance. After all, whatever does happen to a system may be said, thenceforth, to be that system. Systems are hard put not to maintain themselves: If a system changes, those changes afford developmental maintenance of the system; if a system becomes unchanging, that causal dead-end state exhibits self-maintenance neat; and even if the system disintegrates and dissipates its parts throughout the universe, the system has to some extent maintained itself reproductively, for surely those parts have something in common with the system they once comprised. Now any system that behaves differently under different sets of field conditions will have plasticity with respect to its end (or goal) of self-maintenance. And every system may be adjudged to behave differently in different circumstances, insofar as conjectural, subjunctive testimony is admitted. So without the usual lack of surety attendant upon predictions, we may forecast that, given existing field conditions, conditions that necessarily fall within the system's variancy (with respect to self-maintenance), the system will attain its goal of self-maintenance.

It follows that the appeal made to a system's larger end of self-maintenance, loosely conceived, does not restrict teleological explanations to a proper subset of those licensed by Braithwaite's account. To the contrary, in the absence of a nontrivial specification of that end, teleological explanations are, on the basis of such an appeal, applicable to every system whatever.

But instead of trying—by seeking some pertinent, nontrivial specification of self-maintenance—to prop up the proposal to tie teleology to self-maintenance, it might be wise to look for a less Aristotelean role for the notion of Braithwaitean plasticity to play in the explanation of organic activity. One especially promising role, curiously denied by Braithwaite, is for that notion to provide us with a fuller understanding of how things work than might be provided by the details of the one causal mechanism that operates on a given occasion. Braithwaite suggests that the real value of a teleological explanation that attributes plasticity to a system is largely lost in a case where we already happen to know the causal laws involved, "for in this case, that the causal chain which will occur will lead to the goal . . . [will already have] been calculated from its 'mechanism.' " [14] And since the straightforward causal explanation would have been quite satisfactory, the giving of a less informative teleological account based on that explanation strikes Braithwaite as unprofitable and disingenuous.

But not all explanations couched in terms of plasticity need to be less informative. For it is surely worth knowing about an organism not just that it attains each of its eventualities in the way that it does but also that it has the capacity to attain some of them in other ways besides. We may ask for and get a satisfactory answer to how an organism did such and such, but it would be in our scientific interests not to shun a fuller account that tells us more than we asked for, an account of the organism's underlying plasticity. (Besides, if teleological explanations ought to be stated in terms not of Braithwaitean but of many-one plasticity, which requires an additional multiplicity among the causal chains leading to a single "goal" event, then knowledge of those chains is, pace Braithwaite, an essential basis for those explanations. Such knowledge is the only sure way to establish the system's plasticity.)

Now if we are to use the notion of plasticity to arrive at an accounting of how an organic system works under varied circumstances, we will need a prior sense of the functioning whole whose operations we mean thus to explain. That sense is perhaps best provided not by restricting our attention to the organism but by conceiving of it as an integral part of an epigenetic system that includes the environment, too.

Such a conceptualization may be a bit difficult to reconcile with the intuition that the notion of teleology is most suitably applied to just part of the system, to the organism neat; but, if we do decide to retain that notion, applying it to the epigenetic whole will have some mitigating advantages: The fictive case of irradiation-inclined, grotesqueness-prone plants will pose less of a problem. It had seemed unreasonable to suppose that whether or not goal directedness was to be ascribed to one of these plants would depend upon facts about the most common growing conditions for the other plants of the same genotype. And it seems comparably wrongheaded to insist that only of tendencies expressed in "natural" growing conditions might teleological descriptions properly be invoked, for what constitutes the "natural" environment of such artificially mutated plants is moot. However, if the facts of the case are made plain, if the "natural" artificiality of the situation is not concealed, it might reasonably be said of such a plant in situ that it exhibits a teleological tendency toward the grotesque.

But notice that this way of speaking does not seem unalterably opposed to the idea that it is the organism per se that exhibits goal directedness, for although teleology is ascribed to the whole epigenetic system, to the plant in a situation, the plant aspect of that whole is emphasized: the plant not only is given a slight conceptual preeminence over its situation, but the plant also has its proprietary grotesqueness designated as the goal of the system. It is possible, then, that if teleology can reasonably be assigned to whole epigenetic systems, teleology may also be assigned, in some cases, to the organisms that constitute primary subsystems of those wholes.

But how could one ever warrant the ascription of nonintentional goal directedness to an organic system? The earlier appeal to larger goals or ends proved futile in the absence of a nontrivial specification of them, and there is good reason to fear that if goalhood at one level is to be justified on the basis of goalhood at the next higher level, then final justification will need to be deferred endlessly upward. So perhaps the ultimate justification should be sought not at the empyrean heights but on the lowest possible level. Accordingly, one might look to the system's most elemental plastic mechanisms for the key to goalhood. But while there does seem to be some nontrivial end-serving maintenance implicit in the very idea of many-one plasticity, it is stretching the truth to aver that the goal directedness is self-evident in the mechanisms underlying this plasticity. Yet if it is not immediately available within *each* of a system's plastic mechanisms, the desired justification might nevertheless be logically constructed from the class of *all* those mechanisms, where the total number of them exceeds one: Each may then be said to exhibit goal directedness, and all, collectively, said to maintain the functioning system. (This is not to suggest that the functioning of a teleological system is to be equated with the operations of plastic mechanisms alone. In addition to the many-one mechanisms without which it could not be deemed teleological, a given system might also have other mechanisms—though any such nonplastic mechanisms would lack goals. And although nonplastic mechanisms would contribute their functioning to that of the system to which they belonged, they could not be said, nontrivially, to maintain the system or its functioning.) [15*] The demise of a teleological system is its loss of plasticity, its inability any longer to maintain itself.

Of course there is one sense in which this account defers the justification of teleological descriptions—by having their proper application depend upon a sophisticated understanding of the component mechanisms and processes involved in a given system. But if final justification is available only at the horizon of scientific inquiry, it is still reasonable to make tentative use of teleological descriptions much earlier in the course of that inquiry. Here, perhaps, the sort of inductive evidence Braithwaite sees as vindicating such descriptions and explanations might better be seen as improving their credibility, by testifying to the likelihood that the promissory note for their eventual justification will be made good.

Yet even this loosened standard for allowable usage of teleological descriptions can seem overly stringent, since there is an alternate intuition that nonintentional goal directedness is in the eye of the beholder. This alternative is on firm ground as a way of putting the pragmatic point that we assign to the world various categories (here, teleological ones) on the basis of our own interests and purposes. Given our more scientific interests and purposes, there might seem to be no serious opposition between this pragmatic alternative and the proposed standard(s), but there are some cases where the two options conflict. Thus, a wind-up doll, which either works in a particular way or does not work at all, which hence lacks many-one plasticity, might be assigned some measure of goal directedness by the one option but not the other. Even granting the supposition that the primary sense of goals is an intentional one,[16*] I would insist that the extended usage in which goals are attributed to wind-up dolls has gone too far: Thoroughly nonplastic devices and their operations might be said to achieve the goals of others but not themselves. When it is only in the eye of the beholder, goal directedness is not reasonably assigned elsewhere.

5.5 The Conceptual Compatibility between Plasticity and Instinct

Some prime examples of putative instincts are animal behaviors directed upon environmental objects. The behaviors are supposed to exhibit some (genetically predetermined) adaptive know-how concerning these objects—commonly called "goal-objects." But applying here a Braithwaite-type analysis of teleology requires some doing, since that analysis is couched in terms of goals, not goal-objects. An initially more promising analysis proposed by Rosenblueth, Wiener, and Bigelow,[17] *is* expressed in terms of goal-objects (from which the animals and machines receive "negative feedback" and toward which, on the basis of that feedback, their behaviors are guided). But one difficulty, suggested by Richard Taylor,[18] is that, in the absence of any goal-objects, teleological descriptions are ruled out by this analysis. Scheffler points out that Braithwaite's account avoids the difficulty by avoiding mention of goal-objects.

How, then, might the apparent teleology of instincts be analyzed? The earlier proposal that an organic system to which plasticity is attributed should be viewed as an epigenetic whole, as an organism in situ, now serves to facilitate a Braithwaitean interpretation of goal-objects: They are environmental objects internal to the epigenetic system striving to maintain itself. As such, they may (and do) figure in its internal, goal-attaining chains: The chains are causal sequences of events, and goal-objects are among the *things that something happens to* that constitute those events. So, for example, an apple could be a goal-object that figures in an event, someone's eating an apple. In some of the cases where a goal-object figures in a goal event, a negative-feedback model might help explain, mechanistically, just how the system variously attains its goal. Notice, though, that a goal-object need not figure in a goal event per se, only in some event of a goal-directed chain. This is not an oversight: suppose there is a certain goal-object, tripe, toward the eating of which I exhibit, or my digestive system exhibits, no plasticity whatever. Under only one particular circumstance, starvation, and in only one specific way, swallowing without chewing, do I ever eat the stuff. This might nonetheless count as a goal-object insofar as the eating of it is part of a causal sequence terminating in some other event—say, my becoming physiologically satiated—respecting which I do exhibit teleological plasticity.

This neo-Braithwaitean analysis is supposed to include all the revisions proposed in the last two sections. Two of those revisions, the application of teleological descriptions to *whole epigenetic* and to *specially self-maintaining* systems, effectively conspire to blur J. V. Canfield's distinction between models of teleology: a *target* model, exemplified by "a homing torpedo proceeding to its goal" and a *furnace* model, exemplified "by a house equipped with an automatic furnace." [19] The latter revision implies that a system might be said to have its own targets only insofar as their attainment might further the system's self-maintaining, furnacelike functioning; the former revision suggests, further, that such targets are, from the standpoint of the system, some of its internal conditions in want of adjustment, as are temperatures within the house that are lower than those preset on the thermostat.

The resultant perspective is not the usual one of ethological observers who describe putative environmental goal-object-directed instincts; but it does not preclude such descriptions, either. For if, as formerly suggested, the organism could be construed from such a perspective as the principal subsystem of the epigenetic whole, then there is no reason why items external to the subsystem could not be described as its external (or environmental) goal-objects.

Given Kuo's desire to eschew talk of teleology, it is ironic that further analysis of his own notion of plasticity has yielded material for a plausible reconstruction of the notion of teleological behavior. The final irony, though, would be to use this notion to resurrect the notion of instinct, and some remarks by Braithwaite begin to suggest how this might be done: "My knowledge of the conditions under which a swallow will migrate is derived from knowledge about past migrations of swallows and of other migrants, fortified perhaps by general teleological propositions which I accept about the external conditions for self-preservation or the survival of the species, themselves derived inductively from past experience." [20]

The fortifying Braithwaite alludes to is a matter of setting specific, lower-level teleological hypotheses in the context of a "deductive system in which they are deducible from a set of higher-level [teleological and other] hypotheses" [21]—that is, deducible from a general teleological theory that thereby explains them. But even if all this could, as he supposes, be supported inductively, more would seem necessary to warrant the additional hypothesis that the swallow's migration is instinctive behavior—directed, say, toward the goal-object Capistrano.

Let us suppose that we know all the physical details about the swallow's developmental-functional processes and mechanisms and, then, ask ourselves what sort of knowledge this would have to be were it to entitle us to conclude that the migrating was instinctive. Specific propositions about the plasticity of those mechanisms would not seem to afford deductively adequate premises to support that conclusion: Knowing that the swallow had plasticity with respect, say, to perching in Capistrano and/or that Capistrano itself figures in a further chain of events, one chain among others, all of which lead to a particular outcome requisite to the swallow's self-preservation, and so on, would not settle whether the behavior, flying toward Capistrano, was innate or not.

But before deciding that the plasticity of behavior is irrelevant to its innateness, let us return to a problem outstanding in past efforts to make sense out of the hypothesis that some behaviors are instinctive—the problem of characterizing the sort of genetic unfolding that might plausibly be said to lead to instincts. Since ontogenesis obviously involves the incorporation of environmental materials and seems subject to the influence of external stimulation, the idea of pure maturation, which could in principle bring forth innate behaviors, is a pipe dream. And while the idea of a normal environment in which the ontogenesis of behavior proceeds along genetically predetermined lines can face up to this problematic commerce between the developing organism and its environment, this idea does not seem to square with the vast contingency of the world. However, one speculative way out of this quandary has yet to be considered—the idea of characterizing genetic unfolding in terms of the plasticity of developmental processes.

Suppose we distinguish, roughly, between behavior-producing mechanisms and the developmental processes or mechanisms that produce them.[22*] Whether there is any nonarbitrary way to mark the end of the one and the beginning of the other of these two phases of ontogenetic functioning is questionable, but if there is not, my present purposes could be served almost as well by referring simply to early and late phases. Most important is, first, that the two phases extend from the beginning of ontogenesis to the completion of the behavior sequence and, second, that both phases exhibit many-one plasticity; for if these conditions obtain, the hypothesis that the behavior is an instinct seems well secured against many of the objections heretofore raised. That the later, behavior-producing mechanism exhibits plasticity is requisite to the behavior's being aptly teleological; that the earlier, developmental mechanism exhibits plasticity, too, strongly suggests the innateness of the behavior.

The strength of this suggestion rests with the conservatory character of many-one plasticity, its capacity for achieving the same outcome under varied circumstances. Accordingly, if genetic material in situ had such plasticity, it might be construed as a developmental tendency to give rise to the same outcome (be it a gross morphological characteristic or a behavior-producing mechanism) under disparate environmental conditions. This way of characterizing genetic unfolding would thus serve to resurrect the notion of instinct without appealing to either of the implausible, nearly mythological ideas of pure maturation and environmental normalcy.

One acid test for the proposed characterization is whether it can withstand the anti-instinctive element introduced when learning processes intervene in the doubly plastic developmental-functional process. Picture, incompletely,[23*] the underlying learning mechanisms as mechanical devices that are switched on whenever a learning process is called for—say, when goal failure is imminent. Now Lorenz, it may be recalled, contends that if specific learning processes intervene only at certain points along the behavioral sequence, then the learning process might be viewed as consonant with the innateness of that behavior, but that if the learning process is not so neatly intercalated, if there is some "diffuse modifiability" involved, then the learning is not thus consonant. And we might think to apply his line of reasoning not just to the behavioral sequence but also to the developmental process leading to it. Indeed, we might even suggest that diffuse modifiability poses no more of a problem than do the highly specialized learning routines: Suppose that "diffuse modifiability" is to be cashed out theoretically in terms of a general learning device—as opposed, say, to behavior-specific devices associated with specialized learning routines. It might then be argued that the device may be as general as you please and still not pose a fundamental challenge to the hypothesis of a behavior's innateness, provided, say, that this device switches on and off automatically, terminating its diffuse operations as soon as other mechanisms are in a position to do what they do to further the production of the behavior. The organism might still be construed, in teleological terms, as bound and determined by its very genes to produce that behavior.

So it appears that mechanistic learning processes or other impingements of the environment on developmental-behavioral sequences should not necessarily dissuade us from asserting the innateness of certain behaviors, given the double many-one plasticity used to back up that assertion. But what many-one plasticity here appears to give, one-many plasticity may yet take away; for if the conservatism of the former points in the direction of one preordained, original animal nature, with its complement of instincts, then the progressivism of the latter points toward many novelties, many possible natures for humans and brutes. Suppose that a Kuoian epigeneticist manages, by way of minor alterations of a swallow's environment, to create a "behavioral neophenotype," a swallow that persists in hopping and swimming, but never flying, toward Capistrano. Now if a swallow could be made to develop this neophenotypical behavior pattern variously, under more than one altered set of environmental conditions, and that pattern could in turn be produced variously, under many sets of field conditions, then this pattern would be no less qualified to count as the swallow's instinct, as part of its pristine nature, than would the flying that has been preempted.

When one-many plasticity enters into ontogenetic causal chains culminating in behavior, serious objections may be raised against the instinctiveness claimed for that behavior. Learning and other means by which the environment nurtures behavior are, strictly speaking, not the problem for a claim to instinctivity that they appear to be: It is only when they warrant the credible attribution of one-many plasticity *and* move the organism to abandon what appear to be its genetically pre-appointed rounds that these nurturers become truly problematic. If, instead, they facilitate the achievement of behavioral outcomes respecting which the organism exhibits aptly doubled many-one plasticity, then no problem is thereby posed.

But what if, as in the case of our intrepid swallow, hopping and swimming toward Capistrano, there is such doubled many-one plasticity with respect to two mutually incompatible goals—say, hopping onto the Capistrano mission and, alternatively, flying onto it? A staunch defender of instinctivity might see all of this as written in the swallow's genes, and even that arch foe of instinct, Kuo, would seem prepared to concede that there is here an undeniable residuum of genetic predetermination, a preset range of behavioral potentials. But those of us who have learned Aristotle's lesson that one should not expect to find a giant oak within the acorn might wonder whether it is much more reasonable to expect to discern within the genotype a detailed accounting of exactly what will happen to it, of what it will do, if . . .

Although patently compatible with some and arguably compatible with other claims to instinctivity, one-many plasticity may not, in all its forms, be wholly compatible with all such claims. Thus, learning-derived one-many plasticity seems quite capable of leading the organism off in uncharted directions, even if the learning process is itself utterly mechanical (though if it is, then perhaps those directions could in principle be charted—by a very smart, very knowledgeable demon). And if the learning is not so mechanical, if it is intelligent, then perhaps the resultant one-many plasticity is all the more difficult to reconcile with instinct. Indeed, intelligence itself might seem to afford the strongest possible antagonism to instinct, to be the very manifestation of a special form of anti-instinctive one-many plasticity. This, of course, remains to be seen. To that end, let's delve further into the nature of intelligence.

Chapter 6

Intelligent Behavior, Purposiveness, and Consciousness

A well-crafted, philosophically seminal account of intelligent behavior was given by Gilbert Ryle in *The Concept of Mind*, his spirited counterwork against Descartes's Myth—"the dogma of the Ghost in the Machine." Taking the structurally sound core of his account as a basis for my efforts, I shall begin in earnest to construct an alternative to instinct. But my philosophical motives seem quite antithetical to Ryle's, since he sets out to avoid rendering mentalistic interpretations even for intelligent behaviors, whereas I hope to construe all purposive behaviors as presupposing, in one way or another, a special sort of consciousness. The notion of this "mode of consciousness" (to flaunt the Cartesian vernacular) will be vital to my case against instinct and to my corollary proceedings against Rylean and other forms of mechanistic functionalism, but more on that later. For now, by way of introduction to Ryle's remarks on intelligent behavior, I shall note their affinity to Dewey's views on that subject.

6.1 Intelligent Habits and Capacities: From Dewey to Ryle

Dewey insists that there is no incompatibility between the mechanistic character of habit and the flexibility of intelligence, but he does distinguish between flexibly intelligent and rigidly mechanical habits:

> How delicate, prompt, sure and varied are the movements of a violin player or an engraver! How unerringly they phrase every shade of emotion and every turn of an idea! Mechanism is indispensable. If each act has to be consciously searched for at the moment and intentionally performed, execution is painful and the product is clumsy and halting. Nevertheless the difference between the artist and the mere technician is unmistakable. The artist is a masterful technician. The technique or mechanism is fused with thought and feeling. The "mechanical" performer permits the mechanism to dictate the performance. It is absurd to say that the latter exhibits habit and the former not. We are confronted with two kinds of habit, intelligent and routine.[1]

Gilbert Ryle draws a very similar distinction between intelligent and habitual practices: "It is of the essence of merely habitual practices that one performance is a replica of its predecessors. It is of the essence of intelligent practices that one performance is modified by its predecessors. The agent is still learning." [2] Dewey and Ryle plainly agree that the items they distinguish are, in Ryle's words, "second natures or acquired dispositions," [3] but the two authors differ slightly in their understandings of the means whereby those items are acquired. According to Ryle, habits are built up by *drill,* which dispenses with intelligence, while intelligent practices are built up by *training,* which develops intelligent capacities. But Dewey, as is his wont, would find some continuity here: For him, all habits are solutions to problems and, hence, involve a measure of intelligence. Even drill, which requires solving the problem of repeating something, does not fully dispense with intelligence. Yet if the difference between drill and training is for Dewey only one of degree, there is still for him a big difference: Drill leads to a rigid, automatic habit, narrowly *adapted* to one set of circumstances, perhaps, but not adaptive to other, more or less novel circumstances; whereas other educational procedures, ones (including training) that demand more extensive problem solving, more exercise of intelligence, lead to more genuinely *adaptive* habits.[4] So Dewey's accord with Ryle is here almost complete.

Ryle proceeds to give a dispositional account of habits and intelligent capacities, where *brittleness* is given as but one example of a dispositional concept: "The brittleness of glass does not consist in the fact that it is at any given moment actually being slivered. It may be brittle without ever being slivered. To say that it is brittle is to say that if it ever is, or ever had been, struck or strained, it would fly, or have flown, into fragments." [5] But whereas simple habits are viewed as "simple, single-track dispositions, the actualisations of which are nearly uniform," intelligent capacities are construed as "higher-grade dispositions," "the actualisations of which take a wide and perhaps unlimited variety of shapes." [6]

Ryle also suggests that an intelligent performance is not a matter of doing two things—exercising intelligence and performing an action—but of doing one thing *in a certain way*—performing intelligently. This is called, for an obvious reason, an *adverbial* account of intelligence. Combining the last two suggestions, Ryle recommends that intelligent capacities be viewed as multitrack dispositions to behave in intelligent ways. He does not go on to explain the intelligence ascribed to behaviors on the basis of their being the exercises of intelligent capacities, for that would be patently circular. Instead, he suggests that in calling a behavior intelligent we are not simply describing the behavior but also appraising it.

To avoid circularity, Ryle might think to base such appraisals on the idea that what makes a given behavior (say, a speech) intelligent is not something internal to the performer but something about the context in which that performance occurs. Thus, a particularly misogynistic speech might be intelligent behavior in front of a politician's most ardent supporters, male chauvinists, while that same speech might be rather stupid in front of another audience—say, a meeting of the League of Women Voters. Yet exactly the same things could be going on in the politician's "head," understood both figuratively and literally, on each occasion—if, for example, she does not know which of the two audiences she is speaking to, having left local arrangements up to her assistants.

But exception to this idea could be taken, since the intelligence of the politician, call her "Mrs. Schafly," might be called into question equally on both occasions: How could she be so stupid as to leave the identity of her audience in doubt? How could *she* be a misogynist? In answer to the first question, it could be observed that one person's stupidity may be another's "integrity"—a willingness to say what she thinks to anyone. And, in answer to the second question, it could be said that at least Schafly's misogyny is no dumber on one occasion than on another. But these replies are of no help to the idea at issue, for they begin to suggest that intelligence is somewhat context independent. Indeed, a politician with no integrity who failed to identify his or her audience before mouthing racist slogans that he or she did not believe would have behaved just as stupidly in front of a KKK audience as before an antidefamation league; only the stupidity of failing to identify the audience would not have appeared as such in the former case.

More than a present context needs to be considered in appraising the intelligence of a performance, and Ryle himself is quite prepared also to take into account the subtext—including the sort of maneuver or attempt the behavior amounts to, the manner of the attempt's execution, and so on. Given some such apt understanding or interpretation of it, the performance is then to be appraised on the basis of just how well it did the thing or things that were to be done by way of it. But this is not to suggest that intelligent behavior must afford workable solutions to problems. Indeed, a clever solution may not work, a fairly silly solution may. Coming up with an ingenious but inadequate solution is still a matter of doing something well, of doing it intelligently.

The chief thing Ryle seems intent to rule out of consideration in appraising the intelligence of behaviors is whether or not they happen to be effects of specifically mental goings on. He observes that

> theorists and laymen alike constantly construe the adjectives by which we characterize performances as ingenious, wise, methodical, careful, witty, etc. as signaling the occurrence in someone's hidden stream of consciousness of special processes functioning as ghostly harbingers or more specifically as occult causes of the performances so characterized. They postulate an internal shadow-performance to be the real carrier of the intelligence ordinarily ascribed to the overt act, and think that in this way they explain what makes the overt act a manifestation of intelligence. They have described the overt act as an effect of a mental happening, though they stop short, of course, before raising the next question—what makes the

postulated mental happenings manifestations of intelligence and not mental deficiency.[7]

One likely category mistake here would be to confuse appraisals and descriptions, but no such mistake is involved either in affirming that covert performances do take place or even in appraising them. According to Ryle, "It makes no difference in theory if the performances we are appraising are operations executed silently in the agent's head, such as what he does, when duly schooled to it, in theorizing, composing limericks or solving anagrams."[8] Since Ryle has declared his antipathy to mind-body dualism, even some of his most savvy, philosophically sophisticated readers have dismissed such passages as oversights. Thus, J. L. Austin writes, in review of Ryle, "He seems successfully to conceal from himself, at essential moments, both the actual occurrence of numbers of experiences which, it seems obvious, do occur and which he himself, when not immediately concerned to eliminate them, does admit to occur."[9] But no self-concealment is called for, since Ryle's antipathy is reserved for further philosophical positings, for any thought, say, that covert performances would themselves stand in need of prior, more "occult causes" to render them intelligent.

6.2 Somewhere between Episodes and Dispositions

Ryle does not propose a purely dispositional analysis of intelligent behavior. He suggests, to take one special case of such behavior, that arguing intelligently "is to perform one operation in a certain manner or with a certain style or procedure, and the description of this *modus operandi* has to be in terms of such semi-dispositional, semi-episodic epithets as 'alert', 'careful', 'critical', 'ingenious', 'logical', etc."[10] To appreciate fully this suggestion, it is necessary to consider his treatment of what he calls "heed concepts." This is a somewhat loose confederation of concepts that Ryle is content initially to characterize by way of some illustrative examples:

I refer to the concepts of noticing, taking care, attending, applying one's mind, concentrating, putting one's heart into something, thinking what one is doing, alertness, interest, intentness, studying and trying. "Absence of mind" is a phrase sometimes used to signify a condition in which people act or react without heeding what they are doing, or without noticing what is going on. . . . We cannot, without absurdity, describe someone as absent-mindedly pondering, searching, testing, debating, planning, listening or relishing. A man may mutter or fidget absent-mindedly, but if he is calculating or scrutinizing, it is redundant to say that he is paying some heed to what he is doing.[11]

Since he is especially intent upon not multiplying mental processes beyond necessity, Ryle appeals once again to the possibility of an adverbial analysis, a replacement of heed verbs with heed adverbs: "We commonly speak of reading attentively, driving carefully and conning studiously, and this usage has the merit of suggesting that what is being described is one operation with a special character and not two operations executed in different 'places' with a peculiar cable between them." [12]

But Ryle notes that this maneuver does not rid us of the temptation to multiply entities, to speak of something more than an objectively observable manner of doing things:

> A horse may be described as running quickly or slowly, smoothly or jerkily, straight or crookedly, and simple observation or even cinemagraph films enable us to decide in which manner the horse was running. But when a man is described as driving carefully, whistling with concentration or eating absent-mindedly, the special character of his activity seems to elude the observer, the camera and the dictaphone. Perhaps knitted brows, taciturnity and fixity of gaze may be evidence of intentness; but these can be simulated, or they can be purely habitual. In any case, in describing him as applying his mind to his task, we do not mean that this is how he looks and sounds while engaged in it.[13*]

To avoid the metaphysical hazards, Ryle walks a fine line. On one side is a purely dispositional view, which (on his understanding) would take into account only objectively observable behaviors. On the other side is a purely episodic view, which "would be a relapse into the two-worlds legend"[14] he abjures. The first danger is that "heed" will be reduced to nothing; the second, that too much will be made of it—so, for example, "since minding would then be a different activity from the overt activity said to be minded, it would be impossible to explain why that minding could not go on by itself as humming can go on without walking."[15*] The tightwire Ryle hopes to walk is semidispositional, semiepisodic.

Ryle suggests that the substance of his tightwire is a "frame of mind": "While we are certainly saying something dispositional in applying a heed concept to a person, we are certainly also saying something episodic. We are saying that he did what he did in a specific frame of mind, and while the specification of the frame requires mention of ways in which he was able, ready or likely to act and react, his acting in that frame of mind was itself a clockable occurrence."[16]

To explicate this notion of a frame of mind, Ryle introduces the idea of a "mongrel categorical statement"—a statement (such as "You *would* do that") that serves to explain (in terms of a tendency or disposition) why the thing was done and that can serve (by virtue of its implicit hypothetical statement of a tendency) as a basis for predicting further doings, too. Applying this idea to the case of behaving heedfully, Ryle writes:

To say that someone has done something, paying some heed to what he was doing, is not only to say that he was, e.g., ready for any of a variety of associated tasks and tests which might have cropped up but perhaps did not; it is also to say that he was ready for the task with which he actually coped. He was in the mood or frame of mind to do, if required, lots of things which may not have been actually required; and he was, *ipso facto*, in the mood or frame of mind to do at least the one thing which was actually required. Being in that frame of mind, he *would* do the thing he did, as well as, if required, lots of other things none of which he is stated to have done. The description of him as minding what he was doing is just as much an explanatory report of an actual occurrence as a conditional prediction of further occurrences.[17]

There are, of course, frames of mind other than those involved in behaving heedfully, some of which may even interfere with attentiveness—for example, a student's hostile frame of mind may make the teacher's lessons pass unnoticed. And as A. R. White observes, not all uses of attention-concepts in statements yield mongrel categoricals—for example, the statement "He is playing the piano attentively" is not a mongrel categorical: "A man who plays the piano attentively does not manifest his attention by playing . . . ; he does not play because he is attending."[18] It is, according to White, only when heed-concepts are used in statements to explain behavior that the statements are mongrel categoricals. So, let us restrict our attention to heedful frames of mind which are purported to explain behavior.

Ryle elaborates his mongrel-categorical analysis of heeding by drawing an analogy: "When a bird is described as migrating something more episodic is being said than when it is described as a migrant, but something more dispositional is being said than when it is described as flying in the direction of Africa."[19] Migrating explains the objectively observable behavior of flying south but not, Ryle advises, as a cause may explain an effect: "The process of migrating is not a different process from that of flying south; so it is not the cause of the bird's flying south." Migrating is the *same* process (where "process" means "overt behavior") as flying south, but describing it as migrating is to use "law-impregnated" terms: "The verb 'migrate' carries a biological message."[20]

Applying this to a case in which a heed-concept is used to explain behavior, Ryle says: "So, too, when it is asked why a person is reading a certain book, it is often correct to reply 'because he is interested in what he is reading'. Yet being interested in reading the book is not doing or undergoing two things, such that the interest is the cause of the reading. The interest explains the reading in the same general way . . . as the migrating explains the flying south."[21]

Ryle's insistence that reading out of interest is no different qua performance, say, from reading because one has been ordered to do so, springs from his conviction that the chief difference between the two cases resides in what the reader would do in other circumstances. His position is defensible, especially if we understand him to be suggesting that the performances need not be different. Thus, I might be as interested in a book I am ordered to read as in a book I read out of interest; but, if I am not reading out of interest, then my reading may come to an abrupt end as soon as the order is revoked. My reading could be the same on both occasions, even if my dispositions were quite different. But what about a case where, despite the fact that one has been ordered to read something, one finds it interesting enough to continue reading even after the order is revoked? This still squares with Ryle's suggestion, since the case amounts to one in which, despite having been ordered to read, one is also reading out of interest. Actions, as Freud has taught us, may be overly determined.

Since the case of reading because interested involves what we usually take to be some mental goings on, it is not apparent that Ryle's account of heed has left anything out. But consider that account in the absence of an illustration:

> To describe someone as now doing something with some degree of some sort of heed is to say . . . that he is actually meeting a concrete call and so meeting it that he would have met, or will meet, some of whatever other calls of that range might have cropped up, or may crop up. He is in a 'ready' frame of mind, for he both does what he does with readiness to do just that in just this situation and is ready to do some of whatever else he may be called on to do.[22]

Here it may more easily be discerned that his account requires nothing mental to occur, only for there to be *readiness* of an appropriate sort. Even a crude booby trap can be ready to explode if anyone steps on it, yet it could hardly be said to be in the "frame of mind" to do so. Ryle could, of course, object that this bomb's dispositional readiness is not suitably multitrack, that to be said to be behaving heedfully, one has to have the readiness to do a variety of things, each of which would serve to accomplish the doing of the selfsame thing that one is in the "frame of mind" to do. But this last sort of readiness could still be attributed to a mindless mechanical device, provided only that it exhibited many-one ("teleological") plasticity; so there is nothing inherently mental about what Ryle calls a "ready frame of mind." Ryle may fairly be judged to have left mind out of the picture—the picture of the frame. Since a metaphysical purge of mind is presumed to be the main intent of his book, this judgment will seem neither surprising nor a criticism—especially if the only way to avoid that judgment is to reembellish Ryle's refined portrayal with some idle apparition.

6.3 Further Features of Rylean Frames of Mind

U. T. Place has suggested that Ryle's (semi)dispositional analysis of heed would be well supplemented with a state of consciousness, "an internal state of the individual which is a necessary and sufficient condition of the presence of such a disposition." [23] But just what sort of state would this have to be? Place clearly cannot mean to be speaking of the state of being conscious as opposed to being comatose, "knocked out," or under general anesthesia, and so forth, since "being conscious" in this sense would be a necessary but by no means a sufficient condition for the applicability of heed-concepts. Otherwise, for example, a conscious student would be an interested and attentive one, would be one who was alert, studying, concentrating, and so on.

Were we to view consciousness as William James does, as "a succession of states, or waves, or fields . . . that constitute our inner life," [24] then Place might be interpreted to mean that such fields are necessary and sufficient for behaving and being disposed to behave heedfully. But on James's account of them, these fields are always complex: "They contain sensations . . . , memories . . . and thoughts . . . , feelings . . . , desires and aversions, and other emotional conditions, together with determinations of the will, in every variety of permutation and combination." [25] Arguably, then, a particular field might be devoid of any apposite variety of heedfulness.

What might make more sense of Place's suggestion is to interpret (as James also does) "an internal state of consciousness" as a state of being conscious (or aware) *of* something. The strength of the suggestion, thus interpreted, rests on the conceptual affinity between "being aware of something" and "taking heed of something"; but even if both phrases were identical in meaning, Place's suggestion could not be upheld. We are aware of many other things besides our own actions, and even when our actions are what we are specifically aware of, we are not always behaving heedfully in virtue of the awareness—we may even be well aware of behaving with abandon. If we are to be described as behaving heedfully, then our behaviors must surely, as Ryle contends, be accompanied by a multi-track variety of dispositional readiness; but merely being aware of our behaviors does not assure that readiness. Such awareness might possibly be construed as an *actualization* of our dispositional readiness but not as essential to it (any more, say, than actual breaking is essential to fragility).

Now these failed efforts to sustain Place's suggestion should not be taken as favorable to Ryle's semidispositional account of behaving heedfully, for that account may still be accused of failing to do justice to the indispensable role consciousness plays in all truly heedful behavior. Place may be said to be on the right track but to have failed to specify what sort of consciousness is thus indispensable; and the failed efforts to employ received conceptions of consciousness to vindicate Place's suggestion may merely point to the need for a new conception of the form consciousness takes in connection with behaviors rightly deemed heedful, behaviors done in an aptly "ready" frame of mind.

As the phrase is ordinarily used, "frame of mind" refers either to a rather general quality of a person's character or intellect or to a fairly specific mental state, mood, or attitude. But Ryle employs this phrase as a term of art, with a somewhat different meaning. The heedful frame of mind of a man driving carefully brackets only the driving: Reaching his destination, the careful driver might well throw caution to the winds. The careful driver may not be a generally cautious person or even be in a particularly cautious mood while driving cautiously. "Ready" frames of mind might be said to consist of a kind of *preparedness* for certain undertakings, and though they may antedate and postdate the specific behaviors involved in those undertakings, Rylean frames are exactly cotemporaneous with the behavioral dispositions they ground.

By means of the well-chosen example of an obedient soldier, Ryle indicates just how mindless a ready frame of mind can be. The case is said to be that of an "action which, though quite uninventive, involves a degree of heed." [26] The soldier's disposition is multi-track, though, and his militarily obedient behavior could take any of a number of different forms—depending chiefly on what he is ordered to do: "The description of his frame of mind contains a direct reference to his orders and only an oblique, because conditional reference to fixing his bayonet. His action of fixing his bayonet is, so to speak, executed in inverted commas; he does it as the particular thing actually ordered. He would have done something else, had the order been different. He is in the frame of mind to do whatever he is ordered, including fixing his bayonet." [27] About the only cogitation Ryle concedes to this frame of mind is that which the soldier needs to understand his orders and then, blindly, to carry them out: "Fixing his bayonet obediently is certainly fixing his bayonet with, in some sense, the thought that this is what he was told to do. He would not have done it, had the order been different or been misheard, and if asked why he did it, he would unhesitantly reply by referring to the order." [28]

Ryle is sensitive to the issue of the incompleteness of his account of the soldier's fixing his bayonet *as* something ordered, of the soldier's doing it purposely, in order to follow orders. Asking himself how this differs from other cases that seem not to involve acting on a purpose, Ryle says:

> At least a minimal part of the answer is this. To say that a sugar-lump is dissolving, a bird migrating, or a man blinking does not imply [learning to do so]. But to say that a soldier obediently fixed his bayonet, or fixed it in order to defend himself, does imply that he has learned some lessons and not forgotten them. The new recruit, on hearing the order to fix bayonet, or seeing an enemy soldier approaching, does not know what to do with his bayonet, how to do it, or when to do it and when not to do it. He may not even know how to construe or obey orders.[29]

The point of this answer might be cast, somewhat tendentiously, as follows: Ryle wishes to minimize the role of thought in a ready frame of mind, yet he also wishes to distinguish patently *thoughtless* (physical, instinctive, and reflexive) reactions from seemingly *purposive* behaviors. What he needs, then, is an account of purposiveness as minimally mindful. Now if learning can be used to account for why the soldier behaves as he does, namely, obediently, then his (apparently mindful) purposive behavior of following orders might be reinterpreted as more blindly mechanical than thoughtful.

A sense of how this reinterpretation might work can be borrowed from some (not especially Rylean) remarks of Israel Scheffler. He suggests that a "learning interpretation" of some reputedly "teleological" descriptions of behavior might serve to avoid making reference to the philosophically puzzling idea of future (and/or absent) goal-events (and/or goals) that determine present behaviors. In keeping with a "normal causal explanation" (in which causes precede effects), the learning interpretation in question construes past goal-attainments as causing (via learning) present behaviors to be directed, as it were, toward uncertain future goal-attainments of the same type as the earlier ones. And Scheffler further observes that even "machines capable of learning through the effects of their own operations [would be] equally subject to such teleological descriptions." [30]

Now although Scheffler's remarks support the idea that some mechanistic learning processes might give rise to altogether mindless teleological behaviors, it is not obvious that this lesson can be applied to Ryle's attempt to minimize the soldier's mindfulness. Ryle himself would balk at the suggestion that learning could ever be utterly mechanical; indeed, he contends, plausibly enough, that "picking things up by rote without trying to do so is the vanishing point of learning." [31] And given this view of learning as purposive, there seems little possibility of explaining away purposiveness as nothing but an effect of prior learning. Any such reductive ploy seems subject to the sort of regress argument Ryle frequently uses against others: The prior learning, if purposive, would have to be explained away in terms of still earlier learning, which, if purposive, . . .

But if this regress indicates that Ryle could not entirely explain away purposiveness, the option still remains for him to suggest that a learning interpretation might separate the mindfulness presumed inherent in purposes from the soldier's currently ready frame of mind. Thus, the soldier might now be ready to perform only those specific acts of obedience already within his repertoire, acts he once learned *mindfully* but that he is now prepared to execute *mindlessly*. And mindless though the soldier's behavior may be, Ryle variously characterizes the soldier's bayonet fixing as done "for [a] purpose"; [32] as done "on purpose to do what he is told, or on purpose to defend himself"; [33] and as "acting on purpose." [34]

Rylean frames of mind are inseparably bound up with purposes, and the behaviors (whether highly intelligent or "quite uninventive") to which the frames give rise are so intimately related to these purposes as to warrant calling the behaviors "purposive." [35*] Now if all Ryle wanted to suggest were that purposive behavior is not inherently mindful, his position would be at least defensible. However, he seems intent upon the stronger claim that even those purposive behaviors that are *heedful* are not inherently mindful, and that claim is plainly false.

6.4 Introducing *Conscious Readiness*

Picture two different soldiers, one of whom is so occupied in a pleasant pastime that he is managing temporarily to forget the nearby warfare, the other of whom— still at training camp—is waiting expectantly for a drill sergeant's command, and both of whom respond at once, without a moment's thought or hesitation, to an order to fix bayonets. Both soldiers are *ready* to follow the order, but only the latter is quite *consciously* ready to do so. Both may be said to heed the command to fix bayonets, and both may be equally prepared to do so, but both are not equally heedful of doing so. Various modes of consciousness may or may not attend purposive behaviors, but only conscious readiness must attend those purposive behaviors that are heedfully performed. Conscious readiness is a characteristic way of being conscious of what we are doing so as to be prepared to do what is required in the doing of it. The second soldier, consciously waiting for a command, is already acting on the purpose of following orders; but the first soldier, otherwise fully engaged, say, in card playing, is not yet acting on that same purpose.

Conscious readiness should not be confused with another characteristic mode of consciousness pertaining to some purposive behavior—namely, consciousness of the purpose itself. Not all purposive behaviors, not even all heedfully purposive behaviors involve being conscious of purposes per se. Indeed, as Ryle rightly observes, too much consciousness of one's whys and wherefores may actually interfere with one's purposes, by drawing one's attention away from the task itself. A person does not have to pay explicit attention to his or her purposes in order effectively to be guided by them, though sufficiently complex purposive behaviors may sometimes gain from attention paid to their multiple purposes, especially when we are otherwise in danger of neglecting some of them. Such attention can serve to remind one, to reawaken or reactivate one's dispositions; but reminders are not always necessary. Drawing attention to one's purposes is integral to certain psychoanalytic and philosophical endeavors, a shared point of which is conveyed by Socrates' maxim "The unexamined life is not worth living." Serious endeavors of this sort can even modify our purposes and, where unexamined purposes have impeded us, improve our subsequent performances.

A failure to note the difference between the *conscious readiness* to act on a purpose and the *consciousness of* that purpose might be at the root of Ryle's willingness to deny that consciousness is requisite to heedful behavior; for, again, although conscious readiness is requisite, direct consciousness of, explicit regard to, the purpose of that behavior is not. Heedful behavior is *aptly* attentive behavior; behaving heedfully is paying attention to what one is doing in a way that (at least in principle) facilitates one's doing of it. Due heed is, among other things, a matter of being consciously ready, if required, to modify one's performance in order to keep it on track, to have it conform to the purpose of one's behavior. One might say that conscious readiness is the specifically conscious means whereby one's purposes guide one—without one's having exactly to be aware of them or anything else besides what one is doing. Clearly, though, not every way of being conscious of what one is doing is a matter of conscious readiness. Thus, the tightrope walker who marvels at his or her own ongoing performance may, in consequence of that consciousness, fail to keep "on track."

Behaving heedfully involves paying a special attention to what one is doing; and this special attention is conscious readiness. But behaving heedfully further requires one to be *especially* specially attentive, to be *very* consciously ready to keep one's performance in line with one's purpose. This may involve knowing, being aware of, what one is doing, but such knowing does not imply any figuratively inward attentive regard to a purpose per se.

Conscious readiness admits of many degrees. Heedful behavior demands a high degree of conscious readiness; but not all purposive behavior is created heedfully.

6.5 The Role of Conscious Readiness in Purposive Behavior

Following the lead of Braithwaite and others, I shall presume that there is a sharper distinction between teleological and purposive behaviors than is noted in ordinary dictionaries. The difference is supposed to be that the former behaviors are simply goal directed, while the latter are intentionally goal directed; and I shall take this to mean that purposive behaviors are a proper subset of teleological ones. Scheffler suggests that "a standing passenger thrusting his foot outwards suddenly in order to keep his balance in a moving train"[36] affords a plausible case of nonpurposive teleological behavior; still, to the extent that this man might correctly be said also to be intent upon keeping his balance, his behavior should be described as purposive. To be intent upon doing such a thing is to be behaving intentionally and, here, purposively. The man might seem to fall short of having an intention or purpose, but his being thus intent involves conscious readiness, which in turn implies having a purpose. What he may lack, of course, is a consciousness of that purpose.

If one's behavior involves some conscious readiness, then that behavior is safely said to be purposive; but is conscious readiness, of some degree, essential to purposiveness? William James suggests one clear way in which purposive activity might come to involve little of any sort of consciousness—namely, by becoming habitual. When our useful actions have been made habitual, they have been handed over, James says, "to the effortless custody of automatism." [37] Habits, it might be said, allow us to put ourselves on "automatic pilot" and so to perform according to our purposes without much conscious readiness thus to perform. In complex performances subsumed under multiple purposes, various combinations of unconscious mechanisms and conscious readiness may come into play either serially or simultaneously. Thus, in driving a car, we may attend to the road and exhibit relatively *unconscious readiness* to respond to the exigencies of traffic signals, uneven road surfaces, and so on, all the while relying on the unconscious exercise of certain habits involved in our skilled capacity to operate the car; for example, habits of operating the gas, brake, and clutch pedals with our feet.

But can the whole of any purposive activity be relegated to the control of blindly habitual, unconscious readiness? Something akin to this might result, say, from being classically conditioned to perform a complex sequence of responses based on numerous S-R links that, although individually rigid, collectively exhibit teleological many-one plasticity. Were a dog thus conditioned, its resultant behavior might be judged teleological though nonpurposive, but only if the conditioning were understood to be nonpurposive, only if the dog were not, for example, trying to learn. But suppose, instead, that someone deliberately goes about having him- or herself conditioned to do something in a blindly habitual manner. Would that person's later, conditioned behavior, however unconsciously performed, not still qualify as purposive?

Without settling the issue of whether habits can render one's behaviors wholly unconscious, I propose that even if they were so rendered, their being prompted or initiated by a regard to purpose would permit those behaviors to qualify as purposive—unless, of course, the prompting were on the part of someone other than the one doing the behaving, in which case, the behaviors might well be said to answer to certain purposes without being purposive as such. (As I use the term, *purposive* means *having* as opposed to merely *serving* a purpose.)

Of course it may be tempting to make the counterproposal that when one has relegated one's own behavior to the control of blind habit, this behavior also should be said merely to answer to one's own earlier purpose(s), not to be presently purposive. But behaviors do not so easily lose their purposiveness as they do, sometimes, lose their purposes: Suppose I stumble blindly out of bed, dress, fall into the car, and drive off, numbly, to work. I suppose that I might be behaving somewhat like a mindless automaton, but my behavior remains purposive even if—having lost my job because I behaved much the same at my workplace until noon—my behavior no longer has the purpose, the point, it once had. There is some ambiguity here in the notion of a purpose: The proverbial chicken might *purposively* cross the road, with the purpose of doing so, but without any further purpose in doing so. Pointless behaviors are not necessarily lacking in purposiveness: If, for example, something is intentionally done in order to . . . , then that behavior has a purpose that cannot be taken away from it. If something is placed, so to speak, before one, as something one will do, then one's doing of it because it is thus placed is purposive, pointless though it may be.

I would maintain, further, that if a given behavior is purposive, then it is either performed under the auspices of conscious readiness or prompted by a regard to purpose, a direct consciousness of purpose. But this might be taken to mean that although one or the other of these forms is requisite, neither alone is; and I also wish to maintain that the former but not the latter is indispensable to truly purposive behavior. So I should add that the prompting role of consciousness of purpose implies conscious readiness, without which, therefore, no behavior may be deemed purposive. Of course merely being conscious of one's purposes may not prompt one to act (if, e.g., one is already engaged in acting upon them), but if the consciousness of a purpose does prompt, does initiate, behavior in accord with it, then one is thereby consciously implementing the (hence purposive) behavior. Otherwise put, one is actualizing a conscious readiness to behave thus, in accordance with one's conscious purpose.

As a possible illustration, consider the following case: Aware of my hunger and of the presence of an appetizing morsel, a peach, I frame the purpose of laying hands on and eating it. Fully conscious of this purpose, I then behave in certain ways, ways appropriate to my purpose—I reach out, grab the juicy peach, then devour it. Regard to my purpose leads me to act, but not just to act randomly. My regard to the purpose prompts my behavior by virtue of my conscious readiness to behave in ways that answer to my purpose, a conscious readiness that is realized, that comes to fruition, in my actual behavior. My behavior may not succeed in its purpose; but if my behavior is truly prompted by my regard to my purpose, that behavior must be a conscious attempt, however inadequate, to fulfill that purpose—that behavior must amount to the realization of my conscious readiness to behave in order to achieve my purpose.

All behaviors prompted by a regard to purpose (indeed, all purposive behaviors whatever) are, one might say, under the auspices of conscious readiness. But what are the temporal constraints upon such bracketings of behavior by consciously ready frames of mind? Does a behavior persist, long after its initiation, in being under the auspices of conscious readiness—even if that behavior is blindly habitual? Suppose that the habits are exercised repeatedly, over a long period, and that they do continue to serve, to answer to, a general purpose—say, brewing coffee for breakfast. Is my prehabitual regard to purpose still prompting me, via some long-ago conscious readiness, in all the years of coffee brewing that follow?

Understandable resistance to this suggestion could be based on the thought that (purposive) behaviors can truly be said to be under the auspices only of *simultaneous* conscious readiness; but is this requirement not too strong? Suppose I exhibit *conscious* readiness only intermittently during a long, late-night drive to work on the night shift. My behavior might still rightly be said to be bracketed by a consciously ready frame of mind—provided that I am not unconscious and that my sometime largely unconscious readiness to continue driving is such that sudden changes in the driving conditions—for example, a semaphore's turning from a green to a yellow light—would occasion my return to fully conscious readiness.

Still, it might be replied, in order for me to notice those changes and respond appropriately to them, I cannot altogether lack conscious readiness; I cannot ever purposively be driving unconsciously to work. Conscious readiness does admit of degrees, and, so the reply goes, no ongoing purposive behavior is entirely devoid of some degree of that mode of consciousness. But I am inclined to reject at least part of the force of this reply, since it seems fairly evident to me that, even though phases of behavior that lack all conscious readiness might otherwise be deemed nonpurposive, if such phases happen to be integral parts of a larger behavior that is purposive, then those phases deserve, derivatively, to be called purposive, too.

Apropos of the other present matter of contention, I suggest that if in the course of my coffee-making endeavors I am unconsciously disposed consciously to modify my behavior as needed when, for example, there is too little ground coffee left in the canister, then my endeavors persist in being purposive even years after their inception via (prehabitual) conscious readiness. (This is not to say that I would be prepared for all kitchen-environmental contingencies, only for some, for various ones.) On the other hand, if when confronted with the almost empty canister I am nonplussed, or I mechanically continue to scoop the same number of times I usually do, then it begins to seem that my present endeavor has lost the purposiveness had by some of its predecessors.

Conscious readiness or a behavior-specific disposition for it is a (conceptual) presupposition of purposive behavior. Thus, one may be said to be behaving purposively even though one has never exhibited any conscious readiness with respect to the purpose involved, provided that in the very course of that behavior one is disposed to exhibit such consciousness in some exigent circumstances. This proviso calls for more than a general capacity to become consciously ready (since the becoming is tied to a specific behavior in accord with a specific purpose) but less than the actual occurrence of conscious readiness (since such readiness may not prove necessary). The larger point, simply put, is that where there is purposiveness, there has been, is, or would be some conscious readiness available to help to put behavior on track or, if at all possible, to keep it there.

Chapter 7

The Functionalistic Turn of Recent Philosophical Psychology

7.1 Ryle's Credentials as a Functionalist

"Functionalism" is a shared name of two disparate (though arguably interrelated) schools of psychological and philosophical inquiry. The earlier, classical functionalism of James, Dewey, and others has been said to have redeployed in the psychological realm Darwin's "core . . . concept of *function,* an assertion that, as a species evolves, its anatomy is shaped by the requirements of survival."[1] James, for example, urged that "consciousness [has] a teleological function"[2] as "a *fighter for ends,*[3] and that consciousness is a survival-promoting causally efficacious "organ added for the sake of steering a nervous system grown too complex to regulate itself."[4] Contemporary "metaphysical functionalism," which takes its cue from computer science instead of Darwinism, would seem to be a significantly different perspective, for according to Ned Block, "Metaphysical functionalists are not concerned with how mental states account for behavior, but rather with what they *are.* The functionalist answer to 'What are mental states?' is simply that mental states are functional states."[5]

A caricature of the history of these two functionalisms might regard behaviorism as playing a crucial, dialectically antithetical role in the transition between them. Thus, Watson's psychological behaviorism might be said to have displaced classical functionalism, which he condemned as a form of psychological study wrongly reliant upon nonobjective, introspective data. And Ryle's behaviorism, which is arguably a philosophical successor to Watson's, might be said in turn to have been displaced by metaphysical functionalism, which reaffirms the reality of mental states. But although this story might even have been acceptable to some of its principals, the lines are less clearly drawn than it suggests. There is, for example, much truth in Herrnstein and Boring's contention that "Watson, in spite of his own denial, was . . . merely a new kind of functionalist, one who strove to bring to psychology not only the evolutionary concepts of modern biology, but also the objectivity of its methods."[6] And metaphysical functionalists are, despite their own demurs, very akin to Rylean behaviorists.

Of course Ryle himself was not a thoroughgoing metaphysical behaviorist, where this would mean that he construed all apparent references to mental states, activities, and so on as really referring to observable behaviors and/or the dispositions to manifest them. Only if metaphysical behaviorists were prepared to concede the existence of covert, in principle unobservable, behavior might Ryle qualify as one of them. But although he is not one in all things, Ryle is a behaviorist with respect to intelligent *overt* behaviors; and this is not as trivial as it seems, since he thereby denies what a Cartesian might well affirm—namely, that what makes those behaviors intelligent is a purely mental activity behind the scenes. Besides, this partial behaviorism is especially pertinent to the issue of the compatibility between behaviorism and functionalism, since an analysis of intelligent behavior is surely the metaphysical functionalists' strong suit.

Leading with what might be thought their weakest suit, Block recounts what functionalists (and other philosophers of mind) should say to a Platonic query, "What is common to all pains in virtue of which they are pains?": "The functionalist says the something in common is functional, while the physicalist says it is physical (and the behaviorist says it is behavioral). . . . Functionalists can be physicalists in allowing that all the entities (things, states, events, and so on) that exist are physical entities, denying only that what binds certain types of things together is a physical property."[7] By the same token, it would appear that functionalists could be behaviorists in allowing that all psychological entities are behavioral, denying only that the essence of pain is *behavioral* (rather than a *functional equivalence* among all pain behaviors). But Block mentions two particular points of contrast between functionalists and behaviorists, the first of which is that

> while behaviorists defined mental states in terms of stimuli and responses they did not think mental states were *themselves* causes of the responses and effects of the stimuli. Behaviorists took mental states to be "pure dispositions." . . . Functionalists, by contrast, claim it to be an advantage of their account that it "allows experiences to be something real, and so to be the effects of their occasions, and the causes of their manifestations." . . . Armstrong says that "[when I think] it is not simply that I would speak or act if some conditions that are unfulfilled were to be fulfilled. Rylean behaviorism denies this, and so is unsatisfactory."[8]

But, as is evident from Ryle's remarks on mongrel categoricals, he affirms the very thing he is here accused of denying. Ryle's position is that a person engaged in a heedful pursuit such as thinking not only would do other things but also is doing something presently—an episodic something that is either an overt or a covert ("mental") performance. Of course it might be charged that Ryle's semi-dispositional, semiepisodic view of such mental goings on makes him more of a functionalist only by making him less of a behaviorist; but this charge falsely suggests both that a more purely dispositional behaviorism than Ryle's would tend to be nonfunctionalistic and that functional states are akin to episodes.

There is something fishy about Lewis's contention that functionalism, unlike behaviorism, allows experiences to be real effects and causes. For if, in "versions of functionalism . . . couched in terms of the notion of a Turing machine,"[9] mental states are analyzed as "machine table states," and those states are said to obtain in any machine (or, system) of which "the counterfactuals specified by the table are true," then this functionalistic analysis is purely dispositional, and as Block notes, "Strictly speaking, none of the states . . . need cause any of the other states."[10] So, on one standard version of functionalism, mental states are only *per accidens*, not in their abstractly functional nature, *real* events.

The same point also holds true of other—possibly all other—versions of metaphysical functionalism. Consider this version: Hilary Putnam explicates the notion of a functionalistic account in terms of the concept of *functional isomorphism:* "Two systems are functionally isomorphic if there is a correspondence between the states of one and the states of the other that preserves functional relations."[11] Fundamental to the idea of a functionalistic theory, that is, a theory about functionally isomorphic systems, is that its posited ("functional") states are identical in each of the systems to which it applies. But not all functional relations are causal relations (e.g., not fitting into a round hole may be a function of being a square peg of a given size or larger[12*]), so there would seem to be little reason to conclude that functional states are part of the Causal Nexus. Indeed, even if all the functional relations expressly sanctioned by functionalistic theories were causal relations, the functional states (of which those relations were the realizations) would still not, given their abstract nature, be real effects and/or causes.

If they were to be understood as purely dispositional, then Rylean frames of mind could easily be interpreted functionalistically, as functional states of a human "system"; because attributing such dispositions would be on all fours with attributing "machine table states" (or, in the alternate lingo, functional states) to a system. If anything, the fact that Rylean frames are not purely dispositional, that they are also semi-episodic and sometimes even covertly so, should seem more of an obstacle to interpreting them functionalistically, because real episodes are more than abstract functional states. However, a closer look at Ryle's account of some such episodes, the ones involved in intelligent performances, reveals that he construes their episodic quality to be incidental to their essence qua exercises of intelligent capacities: Performances are deemed intelligent, quite apart from whether they are either overt or covert, on the basis of how well they do what they were purposed to do. And this suggests the further idea that various performances, whether overt or covert, might be deemed identical in their specifically intelligent aspects—a fully functionalistic idea whose key ingredients are already contained in passages such as this: "The criteria by which [the reasoner's] arguments are to be adjuged as cogent, clear, relevant and well organised are the same for silent as for declaimed or written ratiocinations. . . . [T]he same qualities of intellect are exercised in both, save that special schooling is required to inculcate the trick of reasoning in silent soliloquy." [13] Since Ryle also holds "that exercises of qualities of mind do not, save *per accidens*, take place 'in the head,' " [14] he would appear to allow that the same (i.e., functionally identical) exercises might be performed overtly as well as covertly. All that remains for Ryle fully to endorse metaphysical functionalism is for him to state some such identity conditions for the specifically intelligent aspects of performances, and his not doing so seems more a matter of indifference than of unpreparedness.

If Ryle is right to say that he is only "harmlessly . . . stigmatized" [15] as advancing a behavioristic position, then it might be said, too, that some (viz., Rylean) behaviorists are metaphysical functionalists. More specifically, it might be said that Rylean frames of mind are functional states that would be identical in all systems that happened to be functionally isomorphic to those persons with (or in) those frames of mind.

This construal still faces a hurdle, though; for according to Block, a "second difference between functionalism and behaviorism is that functionalists emphasize not just the connection between pain and its stimuli and responses, but also its connections to other mental states." [16] However, if the behaviorism in question is Ryle's, with his mongrel-categorical analysis of heed-concepts, then there would seem to be no such point of contrast: Rylean frames of mind can easily be imagined to be interconnected with one another—so, for example, cases of multiple purposes might be supposed to involve a sort of nesting of frames within frames, a quasi-logical ordering of a person's set of dispositions (or frames).

Ryle is part metaphysical behaviorist and part metaphysical functionalist, and the possibility of his being both does not depend upon a lack of overlap between parts: His (dispositionally) behavioristic analysis of intelligent overt performances is coherently combined with a functionalistic construal of them. Of course his functionalism here extends beyond his behaviorism, since he seems to hold that covert performances might also be equivalently intelligent. But it should be noted that his functionalism would not be imperiled by a more thorough behaviorism, one that also disavowed the existence of these covert happenings, for divers overt behaviors might still be functionally equivalent "exercises of intelligent capacities." Contrary to some philosophic opinion, functionalism and Rylean (or dispositional) behaviorism are very compatible viewpoints.

7.2 Artificial Intelligence as a Reductio
ad Absurdum of Functionalism

Ryle's credentials as a functionalist are impeccable but have not been accepted as such, and so his views have done little to set the course of contemporary functionalism. Of greater moment are the views of A. M. Turing: His abstract specification of a computing machine and his provocative proposal for attributing intelligence, indifferently, to such a machine as well as to a person have been major guideposts for the combined AI-theoretic/functionalistic turn of recent philosophical psychology. But even granting that functionalistic analyses of intelligent behavior do have much to recommend them, that they do, for example, mark an advance over the naively Cartesian view of intelligent functioning as a purely mental activity, I tend to regard the Rylean (and/or Wittgensteinian) version as a high point in the development of functionalism and to see the chief value of the Turing-style version as a sign of dangers lurking in any thoroughly functionalistic approach to the philosophy of mind.

Ryle should be thought remiss for failing to anticipate the possibility of computer simulations of intelligent behaviors, but remiss only because that possibility affords something like a reductio ad absurdum of his position. Still, one person's reductio, one person's thorn, is another's sharply focused, intriguing implication: Some machines might be intelligent; we might be machines! Who would have thought that Rylean functionalism could come to this? There is cold comfort in thus ridding ourselves of ghosts.

My argument against the AI-theoretic/functionalistic grain is this: All truly intelligent behavior is purposive; all truly purposive behavior presupposes a (dispositional) conscious readiness to conform that behavior to a purpose; the behavior of machines, regardless of how intelligently it answers to certain purposes, involves no such conscious readiness; therefore, the behavior of machines is not truly intelligent. A related argument depends on a suggestion made much earlier (in section 2.13)—namely, that any truly intelligent problem-solving behavior involves grasping a problem as a problem: The purpose of such behavior is to solve a problem, and the conscious readiness to do so necessarily involves grasping that problem as a problem. Since machines are incapable of problem grasping, they lack the requisite conscious readiness to engage in purposive problem-solving behavior; so, the "problem-solving" behavior of machines is not truly intelligent.

A likely rejoinder is that even if the functioning of machines does not involve, it may still presuppose conscious readiness. Indeed, since I conceded earlier that some human behavior might be purposive even when it does not involve conscious readiness, I may now seem unfair to robots when I disallow the same possibility in their case. But consider what that possibility would amount to. It could be claimed that computer behavior does presuppose some conscious readiness— namely, that which is exhibited by the computer's designers and manufacturers. But this is at best a very weak sort of presupposition. Were computers grown like hothouse flowers (in industrial parks?) only the conscious care of the nouveau nurserykeepers would sustain the presupposition. Were these machines to arise from happenstance, the freakish result of natural forces, the presupposition would collapse. My point is this: any presupposition of conscious readiness in the genesis of computer behavior is a bare presumption based on what we know about the design and manufacture of computers. Were the facts quite different, the presumption (hence the "presupposition") would be without basis.

As matters now stand, the history of computer behavior dovetails with the history of some purposive human behavior—that of the designers, manufacturers, and users of computers; so, of course, in some sense computer behavior is purposive. But whose purposive behavior is it? It may be that of the designer; it is unlikely to be that of the manufacturer per se (since the manufacturer's purpose is likely to have been fulfilled once an order for the computer has been filled); it is probably that of the user (assuming that he or she has to do more than plug the computer into an electrical outlet); but it is most surely not the purposive behavior of the computer.

The behavior of a computer may subserve many purposes, answer to them, but none of those purposes belong to the computer itself—unless it can lay claim to them by a conscious readiness to behave in accord with them. Which are *my* purposive behaviors? The ones I can lay claim to on the basis of my conscious readiness to bring them about. Purposes may be owned without being carried out; purposive behaviors may be owned by those who implement them without their having been initiated by those who do the implementing. Once owned, purposive behaviors cannot be disowned. Computers implement purposive behaviors without owning them.

Although there is sometimes just a fine line between purposive behavior and behavior that only answers a purpose, it is important to preserve this distinction. Computer behaviors may answer certain purposes but are never the computer's own purposive behaviors. When a computer is used to serve our purposes, its behavior may be seen as our own. There need be no contradiction in the claim that a computer's behavior is both purposive and not, for it may be our purposive behavior without its being the computer's.

We may worry, science fictionally, that computers will take on lives of their own, becoming self-regulating, self-perpetuating mechanisms. We may conceive the nightmare that such machines will take over, forcing us to subserve them. Lacking a sure cure for this nightmare, we may take cold comfort in the idea that we should never have to concede to such mechanical masters any real sense of purpose (even that which we might grant to a fly-catching frog). What would force us, logically, to concede such a thing?

Suppose we program a computer to monitor its own functioning and report to us, on cue, its preparedness to implement our data-processing instructions. "Are you ready to follow my instructions?" we ask. The computer responds: "Ready to follow instructions." Even if its response is literally true, we should not be forced to concede conscious readiness to this computer.

But suppose a human being is trained to respond automatically to a sequence of instructions that begins with the shouted interrogative "Ready?" and is followed by the snappy reply "Ready!" Only then is the person set to respond aptly to some other commands—"Run," "Walk," "Fire!" The first reply—"Ready!"—seems to be literally true, to be an avowal of preparedness. Must we concede conscious readiness to this human being? Not if the person is thoroughly robotized, perhaps; but is that a real possibility?

Suppose that a soldier, trained as above, pleads not guilty to a war crime, intentionally killing a two-year-old child. His defense is that when the officer ordered him to fire, he did so without a moment's thought. Even if the soldier admits that, before actually firing his rifle, he recognized the child as being no more than a couple of years old, his defense might wash, morally, provided that his recognition came too late—say, only after he had already been pulling the trigger. But if he had time to stop after espying the child, after seeing whom he was about to shoot, then—unless he was in a confused mental state, "in a daze,"—he has come to own his subsequent behavior. This moral appraisal is based on the idea that no human being can be thoroughly robotized—or, at any rate, that conscious readiness is never wholly suppressed by habit per se.

James, who might be thought to have disagreed, actually claimed that, owing to our neurological plasticity, "we do a thing with difficulty the first time, but soon do it more and more easily, and finally, with sufficient practice, do it *semimechanically* or with *hardly any consciousness at all.*" [17] [emphasis mine] This claim is consistent with the preceding idea, though that idea does not really imply that all human endeavors, however rigidly habitual, are done consciously—that is, under the direct supervision of conscious readiness. I suggest, instead, that even though allowing for (virtually) unconscious performances of purposive behaviors, habituation does not abolish conscious readiness. Even when playing no apparent part in purposive proceedings, conscious readiness is still waiting in the wings. Moreover, the bare fact that some purposive behavior has been rendered habitual is no reason to conclude that this behavior has been deprived of its occurrent (not to mention its dispositional) conscious readiness.

7.3 The Elusive Functioning of Conscious Readiness

Computer simulations of intelligent behaviors may mimic conscious readiness, but there is no reason to suppose that the computer then exhibits anything more than the readiness without the consciousness. And readiness neat is not sufficient to warrant calling the behavior purposive, much less intelligent. Url Lanham tells of a computer expert who is asked whether a computer might be made to function just like the brain: "The answer—'You tell me what the brain does in objective terms, and we'll make a computer that will do it.' " [18] Yet even if the expert lived up to this promise, not enough would have been done. Objective terms might capture overt behavior—including "subvocal speech"—and might even testify indirectly to covert behaviors, but those terms are not adequate to the task of catching hold of *conscious* readiness.

To disabuse us of the antifunctionalistic idea that consciousness would be found wanting in otherwise agreeable computer simulations of intelligent functioning, Putnam asks us to contemplate two different sorts of psychological systems— ones that are constituted of mind stuff (or, as he characterizes them, "good old fashioned souls, operating through pineal glands, perhaps") and ones made of more corporeal stuff, which he describes as "complicated brains." These two sorts of systems are to be imagined as occurring in two different possible worlds, and I suspect that Putnam imagines the second, "brain world," to be our own. He continues:

> Suppose that the souls in the soul world are functionally isomorphic to the brains in the brain world. Is there any more sense to attaching importance to this difference than to the difference between copper wires and some other wires in the computer? Does it matter that the soul people have, so to speak, immaterial brains, and that the brain people have material souls? What matters is the common structure, the theory T of which we are alas, in deep ignorance, and not the hardware, be it ever so ethereal.[19]

But Putnam's imagination has here got the better of him: There is something all too strange (even in the singular realm of philosophical speculation) about his would-be soul-inhabited world, about its underlying conception of the mind. Mind is not a matter of raw material—not even according to Descartes, whose view of mind as substance has plainly, if not fully, informed Putnam's fancy. Descartes did not construe mental substances in such a way that it would make sense, however figuratively, to suppose that they were the ethereal stuff from which mental machine parts might be forged, incorporeal wires drawn, and so on. Conscious-substance, Descartes's thinking thing, is always in at least one "mode"—some mental manner of being whose special nature it is to be directed upon something or other. This directedness could be taken to be the function of each and every Cartesian mental mode, but then it should be duly noted that this function is the very essence of such modes (or specific articulations of conscious-substance). That function could no more be regarded, functionalistically, as a merely possible realization of mental stuff, as something made (to be done) with the raw (im)material of consciousness, than could *being extended* be construed as something that might be made to happen to Cartesian corporeal substance.

Nonetheless, we might suppose that in addition to whatever modality a Cartesian soul has essentially, it also possesses, *per accidens,* the functional states Putnam wants ascribed to it—states identical to those of some (otherworldly) "complicated brain." We could, in good (Cartesian) conscience, go this far to accommodate Putnam's philosophical fantasy. But then, insofar as consciousness were to afford such a nonphysical realization of a functionalistic system, that consciousness would not be functioning mindfully. The functional states of this system would only by courtesy be said to be conscious states, since they would be identical to the states of a functionally isomorphic brain, whose own functional states would only by courtesy be said to be brain states. Once we recognize just how courteous metaphysical functionalism is, we should realize that the abstract states it posits are inherently insubstantial, never either conscious or corporeal. This very abstractness appeals to some philosophical tastes but proves hard to swallow: Any philosophical pabulum that makes (the functioning of) consciousness seem mindless surely has something wrong with it.

7.4 The Strange Fellowship of Turing Machines

Of course it could be proposed that consciousness might be directed upon the states of the functionalistic system it realizes, that souls in Putnam's soul-world might be aware of their own functional states. But if this awareness is attributed on the basis of the character of the system's substance, consciousness, then that awareness should be presumed to be absent in the case of functionally isomorphic brains (in Putnam's brain-world) and, so, to be ineffectual with respect to the functional states upon which it, the awareness, is directed in the soul-system. Accordingly, the proposal in question would not introduce into the soul-system that special form of awareness here known as conscious readiness—a form that is more than a merely idle spectator of functional states.

Alternatively, it might be proposed that the role of conscious readiness is already taken into account in any sufficiently rich, sufficiently complex system functionally isomorphic to one's "system," be it one's conscious soul or one's corporeal brain. And it might then be suggested that there is no further fact of the matter about the actual involvement in any of these functionally isomorphic systems (including cybernetic replicants of us) of conscious readiness, that what is called for is a fairly arbitrary decision about whether to say that systems functionally isomorphic to us also exhibit conscious readiness.

A. M. Turing offers an amusing inducement to make a (related) positive decision. He suggests that the controversial issue of solipsism is usually avoided by adopting "the polite convention that everybody thinks." [20] This convention is based, he intimates, on viva voce diagnoses akin to his own proposed "imitation game" test for the intelligence of computers. This "test" has it that if you cannot tell, after communicating with a computer exclusively by means of teletype, whether you are questioning a computer or a human being, then you ought to attribute the same intelligence to the computer as you would to a fellow human. Turing surmises from all this "that most of those who support the argument [against artificial intelligence] from consciousness could be persuaded to abandon it rather than be forced into a solipsistic position." [21]

This seems a ludicrous piece of emotional blackmail: Call apparently savvy computers conscious (or at any rate intelligent) or be prepared to live a life of hopeless skepticism about other people. A measured emotional reaction to this suggestion is to be less worried about the spectre of solipsism than about the sad spectacle of earnest "conversations" between persons and computers. Including cybernetic systems among our conversational companions, conceding to those systems a conscious readiness to respond aptly, might lead us to set aside our hyperbolic doubts about other people's humanity—but not for any good philosophical or psychological reason and only on pain of having immeasurably cheapened our interpersonal relations. My sympathies lie with human beings and not with functionalistic systems. Nothing about their functioning inclines me to the view that these systems manifest any *conscious* readiness to do anything.

7.5 Isomorphic Oddities

Functionalism's ontic commitments are sometimes clarified by appealing to the type-token distinction—where, for example, the atom is a *type* of entity posited by certain physical theories, while a particular atom whose behavior might be explained and predicted by such theories is a *token*. Metaphysical functionalism's principal commitment might be said to be to equivalent types of functional states of the divers systems, natural or contrived, to which a (psychological) theory correctly applies. This way of putting things seems rather innocuous, especially when combined with the suggestion that the ontic commitment to types is, as in the sciences generally, nothing but a commitment to the tokens of that type. But if this is all that is meant by metaphysical functionalism, then it is not worthy of its name. True metaphysical functionalism should be nonnominalistic to its core: It should not merely draw attention to the fact that divers systems happen to be functionally isomorphic but should also use that fact as a basis for asserting the ontologically irreducible identity of the functional states realized by each of those systems. Psychological-metaphysical functionalism identifies certain types of functional states (i.e., states functionally isomorphic to those of a model of some psychological theory) as abstract structures. These structures are said to be the very essence of psychological reality. Either this reality does not amount to very much or functionalism is incompatible with nominalism.

Another odd feature of functionalism might be made plain by considering the following cases. Suppose that there is a good (Bohr-inspired) theory of atoms as systems of subatomic spheres and that our solar system proves to be functionally isomorphic to the atomic systems that are the intended models of that theory. It might then be suggested, not too implausibly, that the theory in question is at once an atomic and a planetary theory. But consider another case: Suppose that an economic theory admirably explains the workings of our "free-enterprise system" and that this system is functionally isomorphic to an ecological one, which happens to be a model for another, neo-Darwinian theory. Would it be plausible to say that the economic theory is also an ecological-evolutionary theory? We should not be overly impressed by the exact parallel between the two systems, by their functional isomorphism. The causal mechanisms at work in the two systems are, on the face of it, importantly different despite the parallel; so the lure of reductionism, any thought of equating the theories with one another, ought here to be resisted. So, too, with respect to psychological theorizing: The "mere" functional isomorphism of a human being and a cybernetic system does not warrant the conclusion that a good theory about the former system is an altogether apt theory about the latter, much less that the selfsame psychological reality is exemplified by the cybernetic system and its human counterpart. The character of the causal mechanisms operative in a given system is obviously pertinent to whether or not it qualifies as a psychological system, and functionalism here denies the obvious.

Consider the possibility of two parallel villages, physically quite similar and inhabited by identical twins. The villages are located in remote valleys, separated from each other by tall mountains; yet the daily affairs of village B mirror, an instant later, those of village A. Here is a story to account for this odd coincidence. Assume that current neurophysiological thinking has it that certain brain formations constitute what have come to be called, in homage to Ryle, "readiness-sites" but that one dissident, doubtless mad scientist, Bela, believes instead that the neurological bodies in question are really body-mind interfaces, which only he calls "binds." Bela insists that Descartes was on the right track after all but that he mistakenly identified the pineal gland as (the only) one of these binds. Having figured out some of the details about how binds operate, Bela has implemented a mad scheme to demonstrate his neurological and technological achievements: He has constructed the two villages and populated them with the twins. He has also surgically implanted certain devices within the binds of each set of twins. A-devices, implanted in the village A twins, are tiny transmitters; B-devices, implanted in corresponding village B twins, are tiny receivers. A-devices do not interfere with the behavior of any A-twin, but they do transmit each A-twin's bind-impulses to, and thereby determine the behavior of, each corresponding B-twin, whose own bind-impulses are effectively blocked (or jammed) by the B-devices. This explains, then, the mirroring effect.

Now suppose that the more sober-minded readiness-site advocates have a theory, T^r, which, though silent about the modus operandi of readiness sites, does a superb job of explaining and predicting the behavior of the A-twins. T^r posits that readiness-sites are input/output devices that obey certain functional regularities but says nothing about how the devices work. Nevertheless, T^r provides a very thorough understanding of how the rest of the A-twins' psychological systems operate. Since A-twin systems are models of T^r, and B-twin systems are functionally isomorphic to corresponding A-twin systems, a metaphysical functionalist might conclude that both sets of twins exhibit the same intelligent functioning. Confronted with the fact that T^r remains ignorant of the causal mechanisms governing readiness-site bodies, the functionalist might still insist upon the same conclusion, arguing that T^r does explain, at the appropriate level of explanation, what needs explaining about intelligent functioning and that only a benighted reductionist would require further neurophysiological details.

By philosophical hypothesis, however, the functionalist is quite wrong, since a version of Cartesian dualism is correct. Saying will not make it so, of course, but the point is that even if the doctrine of dualism were so, the functionalist would be liable to mislocate psychological reality—that the grounds upon which functionalism is asserted do not rule out a competing claim of dualism.

That functionalism affords a faulty philosophical interpretation of the parallel villages is quite evident in the case of the B-twins, who despite appearances exhibit no genuinely intelligent functioning of their own, relative to their own B-village environment. The moral for the functionalist would seem to be that specific functionalistic hypotheses ought not to be asserted in ignorance of the specific causal mechanisms whereby apparently intelligent functioning is wrought. But this moral is fatal to the metaphysical pretensions of functionalism, whether or not dualism obtains. Bela's own bind (or body-mind) theory, T^{b-m}, may be quite wrong, even though he is able to carry out his grandiose two-village scheme: His transmitters may be sending impulses derived from those purely physical formations, readiness-site bodies, that interact causally only with other purely physical entities. Yet even if T^{b-m} were false, the case of the two villages would still suffice to show that functionalism affords a faulty metaphysical interpretation of psychological reality, since the B-twins would still not be functioning intelligently.

7.6 Some Philosophical Merits of Functionalism

Although eager to make the antifunctionalistic point that the genesis of a system's functioning is linked to its proper psychological interpretation, I do not wish to suggest that metaphysical functionalism is without philosophical merit. It is moot, however, whether there is any appreciable merit in versions of functionalism that propose that functional identity should be predicated of those systems that are functionally isomorphic models of a yet-to-be-discovered psychological theory. T. Block and J. A. Fodor have argued that some such functionalism gives us "too fine[ly] grained" conditions for the functional identity of psychological states: "Thus, for example, if you and I differ *only* in the respect that your most probable response to the pain of stubbing your toe is to say 'damn' and mine is to say 'darn,' it follows that the pain you have when you stub your toe is type-distinct from the pain I have when I stub my toe." [22*] And besides, pains are singularly inappropriate candidates for functionalistic analysis in the first place: Even if pains were to figure in all systems functionally isomorphic to the psychological system of anyone in pain, those pains would surely be more than the abstract structures they might be said to realize. There is more to pain than its purely abstract, functional identity—if such there be.

What functionalism needs, it seems, are identity conditions that are less finely grained and more judiciously invoked. Now although Putnam's version of functionalism may incline us to render functionalistic analyses of every psychological phenomenon covered by the would-be theory T, it does not actually require us to do so. Indeed, insofar as it affords us the wherewithal to identify subportions of the functional structure of the whole system as specific psychological phenomena, that version would also seem to allow us the option of choosing not to make such identifications—in cases where specific parts of the structure do not seem to capture the reality of the psychological phenomena with which those parts are associated. Specific functionalistic analyses might then be given only in cases where, on the basis of careful philosophical consideration, such analyses are deemed altogether fitting and proper. And if this option seems to create the puzzling prospect of a fractured psychological reality in which nonfunctionalistic phenomena interact with functionalistic psychological structures, then it is worth reminding ourselves that functionalism's distinctive posits, the abstract functional structures, are not events and, so, never cause or effect anything, even each other. Rather, it is the specific realizations of these structures that might affect one another causally; and accounting for how this happens, how the overall system operates, is the responsibility of the theory T, not a special puzzle created by (comprehensive or partial) functionalistic analyses of (models of) T.

But notice that the selective use of functionalistic analyses does not necessarily eliminate the sort of difficulty posed by Block and Fodor: functionalism's identity conditions may still prove too finely grained, may imply, for example, that if you infer "not P" from the same premises from which I infer "P is not true," our inferences will be type-distinct from one another. What is needed are somewhat looser identity conditions than those afforded by the notion of functional isomorphism, and these might be furnished by the broader notion of functional equivalence. Thus, one person's saying "damn" in response to a stubbed toe may be functionally equivalent (but not isomorphic) to another's saying "darn," and your inferring "not P" from certain premises may be functionally equivalent (but not isomorphic) to my inferring "P is not true."

Now this retreat from the (arguably premature) rigor imposed by the concept of functional isomorphism not only has the salutary effect of allowing for more flexible applications of functionalistic analysis to systems covered by a future psychological theory T, but also permits us to render such analyses in the absence of any presumption of or about T. And these latter sorts of functionalistic analyses do have a certain undeniable merit, as evidenced by such fundamentally functionalistic suggestions as these: Some overt and covert behaviors are identical qua exercises of intelligence; identical solutions to a problem can be worked out in divers physical media, and some of these varied workings-out might even be said to constitute the selfsame thinking; and so on.[23*]

But although I am inclined to think that these Rylean and Wittgensteinian suggestions have more real merit than do Putnamesque, Turing-inspired versions of functionalism, it is obvious that these suggestions need further philosophical elaboration. Consider, for example, the eminently reasonable application of the functionalistic insight to the case of working out an architectural problem on paper, with physical models, in the actual process of constructing a building. The suggestion that, despite varying "realizations" in diverse media, all this thinking might be the same recommends itself to our philosophical favor quite independently of whether there is an apt psychological theory about such functioning. But questions do arise about the character, exact or otherwise, of the criteria of (functional) identity that are needed to back up that suggestion, and I conjecture that there is always some sort of teleological factor implicit in such criteria. Thus, two different architects might be said to do the same thinking via two different media in certain cases where what is done with each medium is functionally equivalent with respect to a particular architectural purpose: If one architect's activity with drafting equipment can be said to have yielded the same solution to the same architectural problem as another architect's activity with a physical model, a computer simulation of a model, or the real building, then it is possible to suggest that the functional structure of both architects' thinking might have been the same.

Chapter 8

The Possibility of
Teleological Flexibility

8.1 Variously Realized Teleological Schemes

In preparation for a final struggle against instinct, I shall now bring together some philosophical and psychological themes already developed separately. To begin, I propose that Deweyan habits as well as Gestalt-theoretic principles of configuration be viewed as (partial) realizations of certain abstract functional structures. Let us call these structures, which are most fully realized in successfully completed teleological endeavors, "teleological schemes." In keeping with the Gestalt imagery of a melody (executed by the principles of configuration), these schemes might be compared to musical scores; and in some cases these schemes may even be construed, rather literally, as plans or rules in accord with which successful purposive acts are or would be performed. Both Deweyan habits and Gestaltist principles of configuration incorporate concrete causal mechanisms (or, in Gestalt-theoretic terms, dynamisms), so neither of the two can be reduced, ontologically, to the merely abstract teleological schemes. The animate organism (or other psychological system) that possesses those habits or principles may be said to realize the schemes, even before they are actually put into effect by the operation of the (preconfigured) mechanisms. This latter sort of realization is, however, somewhat incomplete—not unlike the reality of a plot that has been hatched but not yet carried out.

The full realization of a teleological scheme is the causal chain of events that, in conformance with that scheme, attains the actual goal of the system, organism, or agent. Incomplete realizations such as those just mentioned might be called "proleptic realizations," to emphasize their anticipatory relationship to fully realized schemes. Still other "narrative realizations," those that occur in discourse (orally, on paper, etc.), need not involve any anticipations of goal attainment. "Trial (or practice) realizations" do attain certain goals, though they are mere prototypes of other goals, schemes for the attainment of which have yet to be (fully) realized. This list of types of realizations may not be exhaustive but does suggest that a given teleological scheme can have a variety of nonisomorphic realizations.

The variety of realizations available to functional structures is commonly deemed important because it affords functionalism an enviable ontological neutrality, an indifference about whether these structures are mentally or physically realized. But the reason now to stress some such variety in connection with teleological schemes has more to do with instinctivity than with dualism: Some possible realizations of these schemes favor instinct-oriented theorizing, whereas others oppose it. This does not make for a quick resolution of the controversy but does create common ground on which opposing sides can join battle.

Even within the category of overtly behavioral, fully realized teleological schemes, there are two major varieties of realization that are directly countenanced by the preceding characterization of terms: The event-chains that realize the schemes may constitute either *intentional* or *nonintentional* teleological activities of a system, organism, or agent. In either case, there will be a prior, proleptic realization of the scheme by the system—a certain readiness (conscious or not) to bring the scheme to fruition. In the case of nonintentional teleological activities, such prolepsis is grounded on the specific causal mechanisms set to realize the schemes fully. In the case of intentional teleological (or purposive) activities, the proleptic realization of a scheme may be partly based on habits poised to realize it fully but must (with or without such habits) be grounded on conscious readiness to implement that scheme.

While it is easy to overestimate the extent to which schemes are proleptically realized before being implemented, it is no exaggeration to say, along with J. L. Austin, that when I do something intentionally "I must be supposed to have *as it were* a plan, an operation-order or something of the kind on which I'm acting, which I am seeking to put into effect, carry out in action: only of course nothing necessarily or, usually, even faintly, so full-blooded as a plan proper." [1] What bears mentioning, though, is that this "operation-order" is almost never, in cases of effective intentional activity, in a finished, final form prior to the undertaking of the activity that carries it (the order) out. Continuous revision, modification, further articulation of such orders, the proleptically realized schemes, in process of implementing them is—as a matter of practical necessity—the rule.

This rule might hold comparably true even in the case of nonintentional teleological behavior, though in this case there is a predetermined character to the processes (of revision, modification, and articulation) and to their final outcome: The system is poised to carry out any of a number of schemes that share the same goal, each of which may be said to be proleptically realized, but only one of which is destined, in existing circumstances, to be implemented. It is quite possible to interpret this move from the many to the one as the selection of one preexistent scheme, even though what is more likely to be involved is an extended process of selection that, after some intermediate winnowing, terminates in a final (proleptically realized) scheme. The motivation for this interpretation, which does further reify the schemes (as existing even prior to their final proleptic realization), is simply to revivify the notion of instinct; for these schemes might, under certain conditions, be construed as innately determined recipes for instinctive behaviors. Indeed, these schemes together with the mechanism(s) for selecting an environmentally apt one among them might be said to incorporate what Lorenz regards as the requisite environmental-informational component of any instinct with full rank.

8.2 Automata, Performers, and Composers

Of course before concluding that a given system's behavior is instinctive, it is well to inquire what that system's relation to its proleptically realized schemes might be—does it mechanically execute them (the way a player piano plays the tunes on its rolls); does it nonmechanically perform them (as a musician might play a store-bought arrangement); or does the system actually devise its own schemes and then implement them (as a composer might first write and then play a work of music)? Let us call these loosely characterized alternatives the Automaton, the Performer, and the Composer, respectively.

A tighter characterization of the first of these options, the Automaton, might be given in terms of a more marked distinction between the "final prolepsis" (i.e., the final proleptic realization of a scheme as implemented and, if successfully, then fully realized) and the original array of "initial prolepses" (i.e., those proleptic realizations of schemes that share a goal and from among which the scheme of the final prolepsis has yet to be selected). Insofar as a system's selection and implementation of a scheme is thoroughly mechanical, that system might be said to execute the scheme automatistically; but the term *Automaton* should be reserved for those teleological systems that not only thus execute a teleological scheme but also thus achieve all their initial prolepses. That is to say, the ontogenetic processes that provide a teleological Automaton with its original array(s) of schemes must be thoroughly mechanistic—devoid, for example, of any intelligent flexibility.

By hypothesis a true Composer does not arrive at its initial prolepses in a thoroughly mechanical way, so a Composer is never an Automaton—though a Composer might execute its invented schemes automatistically. A true Performer does not execute its schemes mechanically, so it too is never an Automaton; but a Performer might also be a Composer of its own schemes. Just as a person may not be a philosopher in all things, so too a system might not be one of the above all the time: A system might be one of the three—Automaton, Performer, Composer— with respect to some schemes while being one of the other two with respect to other schemes. Furthermore, it is quite possible for a scheme to be divided into phases and for a system to bear different relationships to distinct phases. Thus, one might imagine a system that operated as an Automaton with regard to the first phase of a scheme, as a flexible Performer of the second, and then which, as an inventive Composer-cum-Performer, freely improvised the remaining phases.

Instinctivity gains one of its strongest footholds in teleological Automata:

Those teleological schemes (or phases of schemes) whose initial prolepses are produced automatically and autonomously by a given system are leading candidates for nomination as instinctive. This suggests, in keeping with an intuitive understanding of instincts, that any preset programs of a computer-equipped automaton that are not mechanically produced by it during ontogenesis and any programs radio transmitted to a robot (that are not its self-initiated productions) could not qualify as instinctive—even though such programs might subsequently, as schemes, be implemented in much the same, utterly mechanical, manner as those of a true teleological Automaton. The requirement that they develop during ontogenesis does help to capture a naive understanding of instincts, but a more refined sense of them (as inevitably subject to the vicissitudes of environmental influence during ontogenesis) demands added assurance that those very schemes were meant to be, that it was in the nature of the system to bring them in particular into being. Such assurance is forthcoming from the further proviso that the system must exhibit mechanistic many-one plasticity with respect to the ontogenesis of its array of initially realized schemes (together with the wherewithal for selecting and implementing them).

When this developmental plasticity gives rise to functional mechanisms that also exhibit many-one plasticity, that is, that select and implement a scheme from among the many initial prolepses that share one common goal, then (1) the system is a teleological Automaton with respect to that scheme, and (2) the (type of) overt behavior implementing that scheme is an instinct.

As is obvious here (and was already observed in section 5.5), convergent plasticity—especially of the doubled (developmental-functional) many-one variety—is a help rather than a hindrance to instinctivity. But is this plasticity indispensable? In a fictional world of great constancy, ontogenic development might proceed (as it does in the myth of strict genetic unfolding) without hitches along a single, preset pathway; and in such a world a single proleptically realized scheme, the outcome of the aplastic developmental process, might prove workable—adequate to the task of attaining a goal. But although there might not seem to be much practical necessity for a purely automatistic system in that world to exhibit any convergent plasticity, a conceptual need for that system to have multiple many-one plasticity would remain; for without it, the automaton could not reasonably be said, by realizing its scheme in overt behavior, to attain a goal. As proposed earlier (in section 54), nonintentional teleological behavior requires many-one plasticity of the actual processes that give rise to the behavior (i.e., plasticity with respect to the putative goal of that behavior) as well as requiring that the system whose behavior it is be self-maintaining (i.e., that two or more of the system's constituent processes exhibit many-one plasticity). This *multiple* plasticity (the first-order plasticity of the behavior-producing mechanisms and the second-order plasticity afforded by the first-order plasticity of two or more of the system's endogenous processes) does not entail any *doubled* plasticity; but since instinctivity itself requires developmental many-one plasticity, such doubling might still seem inherent in the very idea of nonintentionally teleological instincts. Yet suppose that a self-maintaining Automaton did have (developmental) many-one plasticity with respect to the attainment of just one proleptically realized scheme and lacked any further (functional) plasticity among the processes leading from this initial prolepsis to the behavioral implementation of the scheme. The resulting behavior would qualify as an instinct, for even though there would not have been any actual doubling of many-one plasticity, the developmental plasticity combined with subsequent functional aplasticity could be construed as functional plasticity, too. Perhaps the main drawback of this construal is its not allowing us to identify any one part of the scheme as the Automaton's goal, but were we prepared to accept the whole scheme as a proper goal, that drawback would not be very serious.

Behaviors with goals are, of course, more in keeping with our intuitive paradigm of instincts, so doubled plasticity is very important, if not quite indispensable, to the instinctiveness of nonintentional teleological behavior. The case of intentional goal-directed behavior may be somewhat different, though: Insofar as true purposiveness is compatible with the utter rigidity of those mechanisms that produce behavior deemed purposive, the plasticity of the developmental processes that produce such mechanisms would seem to be all the many-one plasticity required for the instinctiveness of intentional teleological behavior.

8.3 Intelligence as Opposed to Instinct

Having met the enemy, some regular instincts, we must now seek worthy competitors for them. Since the intuitively clearest exemplar of a system with an instinct is an Automaton that has doubled many-one plasticity with respect to a teleological scheme, it might be well to inquire what departures from this pattern would make for an able adversary of instinctivity. I already suggested that a teleological Automaton would be devoid of intelligent flexibility, so a promising departure from the archetype might consist in correcting this deficiency. But what, after all, is so unintelligently inflexible about the instinctive behavior of our Automaton?

The pertinent idea of *unintelligent* (or, to stop mincing words, *stupid*) *inflexibility* implies something more than the rigidity that might be predicated of a teleological scheme itself—more than, for example, excessive regularity or a lack of gracefulness, elegance, or the like. Truly stupid inflexibility suggests a certain unpreparedness or inability to *bend* to circumstance. Simple accord with environmental exigencies is but poor evidence of flexibility, much as the twice-daily correctness of a (possibly stopped) clock is insufficient evidence of its accuracy. More striking, though, is that even a manifest lack of accord in a host of different circumstances may not demonstrate the relevant sort of inflexibility—even when the scheme also exhibits obvious rigidity of its own. Suppose, for example, that a malingering mental patient wishes to persuade his doctors that he is subject to a number of compulsive behaviors, one of which consists of eating in a mechanical way with a spoon—a way so routinized that it rarely serves to bring food (even soup, much less spaghetti or steak) successfully into his mouth. Now despite its manifest inappropriateness in immediate circumstances (as well as its mechanical regularity and gracelessness), this behavior may not be as stupidly inflexible as it appears. It is, perhaps, a slyly intelligent alternative to an otherwise impending doom—say, either a murder conviction that carries a mandatory death penalty or a forced return to a savage military engagement.

Defenders of a dispositional account of intelligent behavior might contend that this case does not pose them any problem, since a larger context would, if duly considered, expose the intelligent pretense for what it is: A malingerer would probably stop behaving oddly once fully out of harm's way. However, a sufficiently cautious malingerer, anxious to save his reputation as well as his skin, might well continue his charade a long time—at least as long as it might plausibly take to be cured by therapy. Moreover, the entire context of circumstances in which malingering serves as an intelligent solution to a life or death problem might be common to a case of truly compulsive behavior, too. So, contextual considerations may not work to differentiate between such cases.

The main problem here for the dispositionalist is not so much that contextual considerations are inconclusive clues to a person's dispositions as that the mere malingerer and the true compulsive could be disposed to behave the same ways. What sets the cases apart from one another is that the malingering is a conscious, willful ruse, whereas the compulsion is more of an automatic response. The malingerer and the genuine compulsive are similarly disposed, yet each has his own disposition in consequence of a different sort of capacity: intelligence vs. automatism.

This appeal to the genesis of dispositions as a basis for distinguishing behaviors that are intelligently flexible from those that are not is quite compatible with the observation that intelligent behaviors are ones that measure up to certain standards. The point is that mere accord with those standards does not, in the absence of the right sort of genesis, qualify the behaviors as intelligent. Thus, a runaway lawn mower that propels an oversized chess piece into its adjoining square, thereby creating a situation known as checkmate, has not done something intelligent; whereas a fairly inept chess player who subsequently calls the move his own has behaved intelligently—even if, left to his own devices, he would have been incapable of discovering that crucial move.

Ned Block makes a related negative point, arguing that "behavior is intelligent only if it is *not* the product of a certain sort of information processing"— a sort involving "very simple processes operating over enormous memories."[2]

His argument uses the hypothetical example of a machine that passes an hour-long Turing test by dint of giving sensible verbal responses to the tester's verbal stimuli. *Sensible* here means that the responses are conversationally apt, and so they may also be said, as behaviors, to measure up to nonetiological standards of intelligence. But Block asks us to consider the possibility of a rather curious etiology for these responses: They are based on the herculean efforts of some programmers, who have listed an enormous though finite set of sensible strings composed of alternating single sentence "turns"—the tester's turn to verbalize, followed by the machine's, and so forth. Block describes how the machine uses this list, let us say, to crib a Turing test. "The interrogator types in sentence *A*. The machine searches its list [on tape] of sensible strings, picking out those that begin with *A*. It then picks one of these *A*-initial strings at random, and types out its second sentence, call it '*B*.' The interrogator types in sentence *C*. The machine searches its list, isolating the strings that start with *A* followed by *B* followed by *C*. It picks out one of these *ABC*-initial strings and types out its fourth sentence, and so on." [3]

Block contends a machine that so operated could pass Turing-style tests but would not thereby have exhibited any intelligence of its own. Could the same not be said of an instinct-equipped Automaton that is capable of passing an ethological equivalent of a Turing test—by responding aptly to environmental exigencies? Notwithstanding any cumulative evolutionary wisdom inherent in its behavioral functioning, a biological instance of such an Automaton might still be thought to produce its instinctive behaviors without recourse to any intelligent capacities of its own. Indeed, the parallel between Block's machine's (search and type-out) information processing and that involved in the selection and implementation of the Automaton's schemes is remarkably close—except that, if anything, there are likely to be fewer items in the latter's "memory," fewer innate teleological schemes with any one goal than sensible conversational strings (in sets of the sort envisioned by Block).

Block cautiously opines that "a minimal degree of 'richness'" is required of the "information processing underlying [truly] intelligent behavior";[4] and the Automaton as described has no richer processing than that of the machine, which according to Block falls short of the minimum. Block further suggests that the relevant "richness" might have "something to do with the application of abstract principles of problem solving, learning, etc."[5] But even if our Automaton-with-instincts were equipped with learning mechanisms that enabled it better to select a fully realizable scheme from among its innate repertoire, the rub is that the lessons to be learned as well as the principles to be applied in the learning of them would have already been wholly prefigured in the initial prolepsis—as integral parts of, or as internally related to, the innate schemes themselves. Comparable remarks would also apply were the Automaton to be equipped with problem-solving mechanisms, for if they did mechanically employ abstract principles simply to solve the problem of selecting a preexistent scheme, the outcome of these machinations would have been foretold among the Automaton's original array of schemes. Indeed, the very procedures used by Block's machine could be said to involve the application of just such problem-solving principles.

Block's appeal to "richness" is almost a tacit commitment to the idea that something no different in kind, only in degree of complexity, is involved in the production of genuinely intelligent (as opposed to his machine's just Turing-test passable) behavior. But even without directly contesting that idea, one may insist that a radically different "processing" would be required to produce behavior that is intelligently flexible, for mere complexity would not alter the fact that no real bending of schemes to circumstance could result from the mechanical playing out of schemes that are preprogrammed into a mechanical system, organic or not.

Mechanistic many-one plasticity of the sort(s) required for an Automaton with instincts not only fails to yield flexibility but even effectively precludes it when mechanistic one-many plasticities intervene within the very processes responsible for that many-one plasticity in the first place: Put somewhat too simply, convergence despite divergences along the way is convergence still. Of course the obvious objection is that one-many plasticity might sometimes be said not just to intervene in but to interfere with mechanistic many-one plasticity.

To begin to assess the prospects for such interference, let us reexamine a type of proleptically realized scheme that was said to have some flexibility built into it—namely, what Dewey calls an intelligent habit. A system's relation to a scheme realized as an intelligently flexible habit is, presumably, more that of a Performer than that of an Automaton; and since a Performer may be supposed to have some intention of implementing the schemes it does, the teleology of its functioning is secure even in the absence of an original array of many proleptically realized schemes. Let us suppose, for simplicity's sake, that a given Performer has but one scheme at its disposal, in its initial prolepsis. An Automaton with instincts of full teleological rank could be said to exhibit mechanistic many-one plasticity not only with respect to (1) an initial prolepsis and (2) the one goal shared by all the schemes in that prolepsis, but also with respect to (3) a final proleptic realization of one among the many schemes in the initial prolepsis. The supposition that the Performer's initial prolepsis has but one scheme effectively precludes the possibility of (2) and (3), though does seem to allow for (1). Let us suppose, further, that the schemes of the Performer's initial and final prolepses have the same goal. This situation is not untypical of cases in which intelligent habits are exercised: The habit's goal is retained even though its original scheme is modified or replaced—the habit's intelligence is exercised in the direction of attaining an end by modifying some means the habit has in its possession.

It is possible, however improbable, for the schemes of an intelligent habit to be identical in the initial and the final prolepses. This could occur if no changes were called for in the process of implementation, and if this did occur the habit might still qualify as flexible. As the Performer has been depicted, its flexibility would consist in the capacity of the habit to modify its teleological scheme without changing its goal. By hypothesis, the habit has no other preexistent schemes to tinker with, so it can only "bend" the one it has. This may be easier said than philosophically understood to be done, however.

Dewey sees no conflict between the flexibility of intelligent habits and what he takes to be the obvious fact that they are mechanisms. "All life operates," he says, "through a mechanism, and the higher the form of life the more complex, sure and flexible the mechanism."[6] If this is to be understood as an expression of determinism, then what he regards as flexibility might best be interpreted as mechanistic (i.e., deterministic) one-many plasticity. But that interpretation puts us in a somewhat familiar bind: What was supposed to be one scheme in the initial prolepsis has been combined with a one-many mechanism (for intelligent processing, learning, or perhaps even for some sort of randomization); yet this suggests that, prior to the final prolepsis of one scheme with goal g, there were many other determinate possibilities the system had for arriving at final prolepses of other schemes, and this is tantamount to the suggestion that all the schemes were initially proleptically realized and that the system does have many-one plasticity with respect both to its final proleptically realized scheme and to the attainment of its goal, g. Mechanistic one-many plasticity has turned our erstwhile Performer into a veritable Automaton—assuming, that is, that the genesis of its initial prolepsis has been mechanistic, too.

Before concluding that instinctivity is thus unavoidable, however, let us consider other means whereby flexibility might be thought to interfere with many-one plasticity. Performers of the sort depicted above (minus the specific hypothesis of mechanistic one-many plasticity) might be construed as Composers of a sort, too: The schemes in their final prolepses are, to some extent, schemes of the Performers' own creation (albeit sometimes a fairly exact recreation of a scheme in a corresponding initial prolepsis). Accordingly, it might be suggested that the Performer's flexibility consists in its capacity to invent novel final proleptically realized schemes. The novelty of such schemes amounts to their deviation from corresponding schemes in initial prolepses, to their lack of functional identity with respect to those prior, presumptively innate schemes.

Alas, the strange logic of "potentiality" is such that it seems all too easy here to attribute to our Performer a functionally identical realization, in the initial prolepsis itself, of the scheme that is realized in the final prolepsis. After all, the argument might go, the Performer at the time of its initial prolepsis could come up with the scheme of its final prolepsis, had the potential to do so, and is that not all that is meant by suggesting that such a scheme is prefigured in an initial prolepsis? Once such an attribution is made, novelty vanishes—and with it goes flexibility. But the argument for this attribution is flawed: something more is meant by a proleptic realization of a scheme than merely that it could subsequently be realized (in a successive prolepsis, in a successful implementation, etc.). What is meant is that the scheme would be subsequently realized if certain conditions were to obtain. A proleptically realized teleological scheme must have a determinate readiness to realize itself another time—a readiness vouchsafed, for example, by causal mechanisms set to bring about the scheme's "second coming" automatically.

The hypothesis of genuinely novel final prolepses patently precludes the possibility that their schemes were previously realized in initial prolepses, but it remains to be seen whether that hypothesis is philosophically coherent. One likely complaint is that the preceding constraints on what is to count as a proleptic realization would seem to be violated by the very realizations used to gauge the novelty of final prolepses: How could those prior realizations be construed as proleptic given that their schemes do not seem bound and determined to be realized again? The strongest answer is that, according to Dewey, those prior realizations, which he calls "intelligent habits," are determinate tendencies toward successive realizations of schemes under the same (or sufficiently similar) circumstances as those in which the habit was first formed. It is only when their schemes prove unworkable that the intelligent habits' capacity for flexibility is exercised, leading to novel schemes (which, when workable, are the makings of newly reconstructed habits). A weaker reply is also possible, though: Initial prolepsis might here amount merely to a likelihood of, or probabilistic propensity toward, subsequent realization of a scheme—that is, the scheme would be likely, though not necessarily bound, to be subsequently realized if certain conditions were to obtain.

But other apparent incoherence remains. What, we might inquire, is the source of the novelty of a scheme in a final prolepsis? A pure Composer, one whose compositions are completely unprecedented, is all too wondrous. Its creations ex nihilo demand divine powers beyond our philosophical ken. Flexibility seems a power of more modest, human proportions; and, even if we do not fully understand it, flexibility seems to be an everyday occurrence that should be reckoned among the very data to be comprehended by our philosophic if not scientific inquiries. Yet by interpreting flexibility in terms of a Performer-as-Composer hypothesis, I may have rendered the ordinary most extraordinary and incomprehensible. Of course it might be observed that this hypothesis is at least less mystifying than the pure Composer hypothesis, which has no prior schemes to guide its inventiveness; but such guidance may not help to explain what most needs explaining—namely, where the departures from the prior schemes come from, the departures that constitute the novelty of the compositions. And although simple flexibility may seem intuitively easier to grasp than do more complicated sorts of inventiveness, it is arguable, for example, that what Dewey calls deliberation is far less puzzling albeit more complex than a unitary intelligent habit's exercise of its flexibility.[7*] The former involves merely recombining old ingredients, from two or more existing schemes, while the latter, which has but one scheme to work with, apparently involves getting something new out of nothing in particular.

Yet such recombinations are themselves unprecedented (or novel), so deliberation is, perhaps, every bit as puzzling as the simplest flexibility. On the other hand, a fairly simple sort of "information processing" might suffice to yield recombinations from preexisting parts of prior schemes. Is anything comparable possible in the case of simple flexibility? Well, if an intelligent habit's sole teleological scheme is composed of separable parts, distinct phases, then there is no reason why comparably simple processing could not serve, for example, to modify the order of these parts or, perhaps by repeating some of them, to extend the duration of certain phases.

This new appeal to "information processing" may suggest that flexibility is more amenable to mechanistic simulation than previously supposed, even that intelligent functioning is thus amenable, too. Still, such artificial intelligence would, for now quite familiar reasons, not afford genuine flexibility. Instinctivity has little to fear from such counterfeit opponents and could even swell its ranks with them—insofar as these processes are innately determined and deterministic. The difficulty of trying to make philosophical sense of the idea of flexibility is becoming a real dilemma: Efforts to avoid the seemingly incoherent idea of novelties from nowhere would appear to succeed only on pain of allowing for the attribution of prior, proleptic realizations of those novelties, thereby destroying the novelty—and with it, any hope for genuine, anti-instinctive flexibility.

8.4 The Phenomenon of Intentionality

A useful preliminary to any further attempt to render true flexibility intelligible is to give some accounting of how a Performer is actually guided by the teleological scheme of an intelligent habit. An intelligent use of such a scheme might be to follow it not too religiously, in order thereby better to attain its goal. From the perspective of the Performer, though, it is not so clear that one is aware of the scheme per se, much less of the guidance it affords. We are not typically cognizant of the personal precedents for our intelligent habits, nor of their schematic contents. The habits and the schemes are known mainly by their effects, the ways they secretly enable us to perform our actions. And it should come as no surprise that we Performers may be guided by, may flexibly conform to, a scheme already mastered but not presently consulted, even in memory—after all, in like fashion one can follow a score without scanning it.

But I do not want to exaggerate our lack of cognizance of the schemes that guide us—or, better, the schemes by which we guide our actions—since doing so might make us falsely appear to suffer our own actions unwittingly. In truth, of course, we palpably do control those actions, in some accord with the schemes, and we are aware of what we are doing as so acting. All I should stress is that there is commonly (and especially in cases of the exercise of simple intelligent habits) a lack of express awareness of the scheme as an overall plan to be carried out, step by step. Insofar as the schemes might be regarded as rules of action, they are ones mastered to such an extent that they enable us to focus more on our doing than the rule following inherent in that doing.

J. L. Austin describes phenomenologically, without his usual recourse to philosophical considerations about ordinary language, some salient features of that focus: "Although we have this notion of my idea of what I'm doing—and indeed we have as a general rule such an idea, as it were a miner's lamp on our forehead which illuminates always just so far ahead as we go along—it is not to be supposed that there are any precise rules about the extent and degree of illumination it sheds." [8] I should like to add that this idea, this conscious sense, of what one is doing is informed by one's teleological scheme, and a rough general rule is that, in the exercise of a simple flexible habit, the illumination provided by the idea reaches, if only in thought, the goal of one's scheme.

To illustrate such an idea, let us return to the case of a lowly frog presumed intent upon catching a fly. Conceding to this amphibian an intelligent (or, at any rate, genuinely flexible) habit of fly catching, we might note that its teleological scheme has a goal that requires a goal-object, a fly, if the scheme is to be fully realized. Let us suppose that the habit is actualized as if in response to the presence, within the frog's purview, of a fly. What, then, is the frog's own sense of that fly? At the risk of putting words into its head, let us say the frog thinks, "Want fly, to eat." This idea of the fly as a goal-object is based on the frog's teleological scheme, but I dare say that the fly is not so self-reflective as to be aware of that scheme itself. Still, I should like to attribute to the frog some non-deliberative consciousness of the fly as *etwas zu essen,* some idea of it, indeed, as that toward which the frog's behavior is self-directed.

Partly comparable attributions might be made in far more familiar cases involving ourselves, thereby avoiding the strangeness of my speculations about other, amphibian minds. Consider the following episode: Ravenous for food, I consult a restaurant menu, call the waiter over to my table, and say, "I want a steak, cooked rare." Now I may have done some thinking about what I want—food, as soon as possible. I may, for example, have contemplated steak tartar but rejected that item on the grounds that although they would not have to cook that stuff, for all I know the chef might spend forever chopping a steak into tiny pieces. But my words to the waiter function for me mainly as instrumental behavior, to get food right away. I am aware of what I say, which might be said to express a thought concerning my intentions, yet I pay attention to what I say not in its own terms but only as a way of getting what I want. Even as I speak the word "steak," though, I think ahead to the juicy object of my desire. I am somewhat attuned to the special exigencies of the social environment and may, to smooth over the waiter's possible annoyance at my brusqueness (an annoyance that might lead to certain avoidable delays), add the magic word "please." My verbal actions themselves, for all that, are compatible with little in the way of conscious deliberation. I do what needs doing to get that food; I puzzle not at all.

The lesson I wish to draw is that even in cases where language indicative of certain goals is manifest, where my very words do correctly convey the nature of a goal-object I mean to obtain, I need not be taken thereby to have focused attentive regard on my teleological scheme. My eye is on the prize, not on the scheme for obtaining it. My lamp, as it were, leaves that scheme in darkness; my idea of its goal leaves the scheme itself unilluminated. The scheme *is* the idea, and though I could pay explicit attention to it, as I do in cases of conscious deliberation about it, I need not do so.

The option of deliberating does not exist for the frog, who lacks the requisite wherewithal, I would say, for self-consciousness about its own schemes. But must we also say that the frog lacks schemes qua ideas? that its teleological maneuvers are devoid of any sense, on the frog's part, of what it is up to in producing them?—to say this would be to deny that the frog's behavior is ever really intentional.

8.5 Brute Thoughts

The frog, unlike the ravenous restaurant patron, lacks the capacity to do things with words, so any idea had by the frog of its goal object could hardly be based on its understanding of its own verbal command. And since the frog lacks even this fairly primitive linguistic capability, we would be hard pressed to suggest that the frog's idea derived instead from one of its own inwardly expressed nonimperative mood sentences, to the covert statement "want food, to eat." Accordingly, all philosophical theorists who regard thought as inseparable from language would deny the very possibility that the frog, a mere brute, had any ideas whatever. Dewey's denial of this possibility is of particular interest, though, since it is against the background of his views that I am attempting to affirm that brutes who have genuinely flexible habits also have ideas of what they are doing as they exercise those habits.

Dewey says that the (anoetic) consciousness of sentient brutes makes them only dimly aware of sensations, not of their meanings qua means to an end. Since the frog's sensations of the fly would have a meaning as an end, a goal-object, my proposal that the frog is aware of this meaning might seem to evade Dewey's stricture. But since that end, the target of the frog's ("preparatory") tongue-lashing behavior is also a means to the frog's ("consummatory") swallowing behavior, the idea I attribute to the frog violates the stricture after all.

I am claiming, in effect, that for the frog the fly signifies something to eat, yet Dewey denies the possibility of signification in the absence of language. Sentient brutes, he suggests, are too immersed in their activity to have a significant sense of its phases, a sense allowing for forethought and deliberation. The frog's putative idea of its goal, fly eating, would not seem to require deliberation but does involve or amount to forethought. And Dewey would insist that forethought requires the same sort of detachment as deliberation does—a sort empossibled by the attachment of meanings to separate signs (which, in the case of silent thoughts, might be "uttered" covertly). His position seems to be that only if meanings come to be carried by sign-vehicles independent of one's actions might one gain (via those signs) a sufficiently independent perspective on those actions to have forethought concerning them.

In defense of his own denial of brute thoughts, Dewey observes:

> The evidence usually adduced in support of the proposition that lower animals, animals without language, think, turns out, when examined, to be evidence that when men, organisms with power of social discourse, think, they do so with the organs of adaption used by lower animals, and thus largely repeat in imagination schemes of overt animal action. But to argue from this fact to the conclusion that animals think is like concluding that because every tool, say a plow, originated from some pre-existing natural production, say a crooked root or forked branch, the latter was inherently and antecedently engaged in plowing.[9]

The "organs of adaption" mentioned are some of the very "schemes of overt animal action" for which I employ the (more general) term *teleological schemes*. But while he concedes that brutes have such schemes, the meanings of their actions, Dewey denies that brutes have the knack to use these schemes in thinking. What is wanted, presumably, are sign-vehicles that are sufficiently independent of an action to be used as props in the dramatic rehearsal of that action and, of course, the requisite imagination so to use them.

Yet not all thinking is a complete dramatic rehearsal. Indeed, the frog's putative idea is no rehearsal at all; so perhaps that particular idea, a simple forethought about the frog's goal, does not require much imagination—not as much, say, as conscious deliberation surely would. The frog still does require a sign-vehicle, but what is wrong, for the purpose, with the fly itself? Dewey might object that an action's goal-object is patently not independent of that action, but it is not obvious that, given the character of the meaning to be assigned to it, this candidate sign-vehicle, the fly, ought to be thus independent: It is this fly as integral to the intended action that is supposed to carry the meaning in question.

One's goal-object (whatever it or one happens to be) has meaning affixed to it whenever one's intent, the idea of what one is doing, is focused on that object as the target of that doing. Dewey would concede that the frog is moved to act by its sentient apprehension (or "feelings") of the fly, even that those feelings are "suffused with the consummatory tone" of the relief from hunger that other flies have given. Still, he questions whether the frog is cognizant of what those feelings mean, is aware of the fly's significance as something to make tongue contact with, to swallow, and thereby to relieve hunger. Although I agree that all this is open to question, I also sanction the possibility that the frog's little, green frame might accommodate some such consciousness of goal-objects; and I see no good reason for denying this possibility in the bare fact that frogs lack membership in a linguistic community.

But perhaps my dispute with Dewey is illusory, since he seems to balk mainly at the possibility that brutes might have ideas suitable for conscious deliberation, where these would be awarenesses of meanings (or schemes). Such awarenesses might arguably depend on action-independent sign-vehicles: The deliberator might well be made aware of meanings themselves only via such signs—though this might prove hard to establish, given that sign-vehicles might be covert and that the deliberator might not even be aware of these vehicles, only of their meanings via them. The ideas I wish to attribute to brutes are somewhat different and might occur even if awarenesses of schemes (e.g., a conscious regard to purpose) did not. My frog might be said not to be aware of its scheme via its goal-object but instead to be aware of its goal-object via its scheme. My frog's forethought about a present fly is thought for food, but may not be food for conscious deliberation.

Further Deweyan opposition to brute forethoughts might also be anticipated on the basis of his dynamic-breakdown theory of consciousness, however, for if consciousness only appears pursuant to a breakdown of habit, then how could a frog just confronted with a fly and not yet lashing out at it already consciously intend to obtain it? The theory's answer would seem to be that the frog could not have such an idea, and it is not implausible to suggest that the frog might not have it: The frog sitting on a lily pad, looking like a hung-over roué, might rather mindlessly lash out with its tongue at a passing fly, consume the morsel, and yet seem barely stirred to consciousness. But would that same theory not make it difficult to deny that in other circumstances the frog would, in consequence of a failed attempt, become acutely conscious of a fly beyond its customary striking range? If so, then that theory proves a better friend than enemy to some attributions to brutes of ideas. But does the theory warrant the specific attribution to the frog of a consciousness of the fly as a goal-object? Since the situation is supposed to have become problematic, and fly catching is, after all, the problem, I am not sure what else the frog might be supposed to have become conscious of besides the fly as something to catch; so I am inclined to take the theory as a warrant of the right sort for the idea I wish to attribute.

But, for all that, I am not so inclined to accept the theory itself, for it would appear generally too restrictive about the occasions in which consciousness is said to occur. In particular, the theory denies the plausible contention that a frog might be quite intent upon catching a particular fly even though not having recently failed to catch it or any other fly. Picture, instead of the indifferent roué, a frog with a lean and hungry look, who has not even seen a fly in days. Now even if the watchful, steady gaze of this creature does not make you want to attribute to it anything more than a certain dispositional readiness to strike out if a fly comes by, would the frog's haste to do so then, maybe even its rapid succession of strikes in the general direction of a too distant fly, not suggest to you the hypothesis that the frog keenly regards the fly as a goal-object?

Frogs aside, Dewey's dynamic-breakdown theory can seem to depict human beings as virtual automatons, devoid of consciousness save when an obstacle to purely mechanical functioning transpires, breathing temporary mental life along the dynamic strands of preexisting habit. This image of something like an impassive spider at the center of its web is difficult to reconcile with our own sense of our mental lives, as more continuously active; so it might help to remember that on Dewey's view we creatures of habit are ever on the move, doing things, unless held in check by conflicting habits and impulses or impeded by environmental obstacles. But to indolent souls like me this reminder compounds the difficulty: A rapidly whirling wind-up doll that thinks when stopped but not before is too mindless and busy for me to recognize me in it.

"Consciousness, an idea," by Dewey's definition, "is that phase of a system of meanings which at any given time is undergoing redirection, transitive transformation." [10] But this definition, which is meant to be construed in terms of the dynamic-breakdown theory, may be too substantive to qualify as a purely philosophical account—one that might promote more immediate understanding, one whose merits might be assessed without recourse to the distinctive methods of science. [11*] To purify the account, let us purge the theory of its quasi-scientific causal requirements. The philosophical core of the theory is the observation that what we are conscious *of* are problems or difficulties, and the validity of this observation does not require that these topics *for* consciousness be causally efficacious in eliciting it.

My interest in this observation is here limited to its application to our consciousness or idea of our own actions. Given Dewey's (purified) definition-cum-theory, such an idea might be said to illumine only those problematic features of our actions that demand the revision—the redirection or transformation—of our schemes for these actions. This might explain why we are not much aware of any sensory qualities or kinesthetic sensations requisite for the control of our movements, since these feelings might not call for us to revise our schemes; but the view would also "explain"—contrary to fact—that we could not be aware of *unproblematic* "ends-in-view" (as Dewey calls the *problematic* variants of them). But how else should we explain the fact that ordinary actions, as Austin says, "involve *incidentally* all kinds of minutiae of, at the least, bodily movements, and often many other things . . . [that] will be below the level of any intention . . . [we] may have formed" [12]? Perhaps the explanation is simply that our focus of awareness is, typically, elsewhere, so these other things do not occupy our attention; perhaps it is even, sometimes, because we are occupied by an unproblematic end-in-view that we are not much aware of all our means of obtaining it.

Action directed toward something as a goal-object, an intended target, does generally involve paying more attention to that something than to the incidental but indispensable minutiae involved. The frog looks to the fly, not to sensory qualities of its own tongue. The expert archer pays less attention to her holding of the bow, thereby enabling her to pay more attention to her target. Such phenomena are mother's milk to Dewey's philosophy, but his dynamic-breakdown theory renders them far too automatic. The less causal, more philosophical reconstrual of that theory allows the agent more leeway to attend to other facets of his/her action, but only insofar as they happen to be problematic. This constraint is, arguably, not nearly so onerous as it seems, since any end-in-view might be said to be somewhat problematic, to involve the problem of attaining it, and as attention shifts to other facets of an action, they tend to become ends-in-view. Yet if it is in the nature of a Deweyan problem to require revision of a scheme of action, then the constraint is every bit as oppressive as it appears: Hitting the bull's-eye is then no Deweyan problem for the expert archer, so she could not pay attention to that target even if she wanted to.

I am in the position of wishing to suggest that consciousness of one's actions is tantamount to their flexibility and yet to insist that Dewey's theory, which ties the two together so closely that they begin to seem as one, is not altogether on the right track. One place that I suppose him to veer off in the wrong direction is where especial care is taken in performing an action. Austin observes that some care is commonly required when one does a thing intentionally: "Whatever I am doing is being done and to be done amidst a background of *circumstances* (including of course activities by other agents). This is what necessitates *care*, to ward off impingements, upsets, accidents." [13]

Dewey might be said to agree verbally that circumstances necessitate care, but to mean by this that care is elicited in direct response to problematic circumstances, that they compel the agent to care. As if to stress that the agent is under no strict compulsion, I would contend that care is voluntary. When we are chided to be more careful next time, it might be thought that our critic is saying something quite compatible with the dynamic-breakdown theory, though advice is usually presumed to be takeable, not to tell us what we are compelled to do anyway. And do we not sometimes chide ourselves for being careless, as if to say that we should and could have exercised greater care in those same circumstances?

Care and caution do not presuppose prior upset. Their reason for being is, rather, to prevent a breakdown. Sometimes we may be encouraged to exhibit more care next time even though our schemes have been fully realized without any revisions this time—for example, when we just barely scrape by, almost failing to realize our scheme fully. But even the master marksman may, should he choose to, exercise special care—because the competition is stiff, the stakes are high, or, even, the whim strikes him. We exercise care, we take heed of what we do, somewhat voluntarily and may shift this attentive regard anywhere along the items encompassed by our teleological scheme as it unfolds, as we unfold it.

8.6 Stretching toward Goal-Objects

A "commodious vicus of recirculation" brings us back to "heed," to consciousness and its executive functioning: Care is predicated on what I have called "conscious readiness," on paying attention to what one is doing in order to do it; but care is more besides, it is a matter of monitoring one's undertaking to make quite sure that the deed is done. When it takes the form of care, conscious readiness might be thought to be a rigid adherent to a scheme, but deeds are sometimes only doable if schemes for them are modified. Conscious readiness in all its guises is none other than the Performer. Conscious readiness is the "onlie begetter" of teleological flexibility. The hungry frog, eager to strike, consciously ready to do so, is presently disposed to do what it can to lay its tongue on the fly, prepared even to bend its scheme of action to problematic circumstances such as, perhaps, a fly too far away. An impassive, satiated frog may lash out indifferently at what comes by, though if its readiness to do so is blind, then it will not concurrently be prepared to bend.

The mise-en-scène of action is brought to life by the actor's conscious readiness to let the play begin. The scheme is first proleptically realized by this consciousness, which anticipates some of the forthcoming plot developments. The mechanical Performer may operate from habit with only the dimmest sense of what he is doing, but will, in consequence, be ill prepared to cope with faulty cues, misplaced props, or other irregular circumstances. The alert, careful Performer will be better prepared, should need arise, to improvise a bit, to depart from some parts of the script in the interest of saving the whole play from disaster— though just how well prepared and able he will be depends on his inventiveness, his intelligent flexibility.

Flexibility is gauged by the novelty of the final prolepsis of the teleological scheme (as actually performed) in comparison with its initial prolepsis (at the inception of the performance). The intelligence of the exercised flexibility demands further assessment on the basis of what the Performer meant to accomplish by deviating from the original scheme.

Conscious readiness empossibles the flexibility of animal behavior, be it human or brute. The frog's awareness of the fly as a goal-object is not divorced from the frog's ongoing action, its purposive behavior directed toward the fly, so the frog cannot use that awareness to deliberate about what it does. But deliberation is not the only means of liberation from the tyranny of preestablished teleological schemes. Some nondeliberative awareness, our conscious readiness with respect to the implementation of a scheme, already grants us some limited freedom from it. A frog, thus aware, is no longer pushed by unconscious mechanisms or dynamisms but now drawn to—though not exactly by—the fly. The alternative to being pushed is not just being pulled; it is also possible to be a self-initiator of behavior that departs modestly from prior innate and/or habitual patterns. A fly does not compel a frog that is consciously ready to obtain it, not the way a stimulus might be said to compel a response. The frog's behavior of striving for contact with its conscious target is not reactive but proactive.

One may of course be aware of items of experience as goal-objects without thereby being consciously ready to pursue them, but the converse is not possible. Conscious readiness with respect to a goal-object is a special sort of awareness of it as a target toward which one is either presently prepared, given an opportunity, to direct one's strivings *or* presently directing one's strivings. This awareness is not an idle spectator of ongoing behavior but part and parcel of the goal-object directed purposive behavior (or action). By itself, the mechanism that propels behavior toward its goal-object seems predetermined to succeed or to fail, depending upon the circumstances. By itself, awareness of something as a goal-object has been free to wander ahead of any behavior actually directed toward that object but seems powerless to implement that behavior. Combined together, as conscious readiness, the mechanism and the awareness are of great assist to one another. If, for example, a previous attempt to "hit the target" has failed, then the conscious readiness to try again may overshoot the range of the prior behavior that fell short of its goal-object and grasp it. Having done so, this consciousness brings actual behavior in its wake. The new behavior goes beyond its precedent's mark, setting a new record. There is no unfathomable mystery in this achievement, which may for all that fall short of the goal-object. The means that made the preceding possible have been modified by the anticipatory, substitute "action" of consciousness; the scheme has been revised by the conscious readiness to do more than had been done before. Conscious readiness has deemed it necessary to expend greater effort, the extent of which is controlled by the consciousness of the goal-object, a consciousness here informed by the prior behavior that fell short, and the direction of which was already set by the original teleological scheme. This feat is a sort of stretching, eloquent testimony to the flexibility of the intelligent habit that made everything happen.

This general case has characteristics not invariably associated with the exercise of flexibility, so no claim can be made that conscious readiness always secures flexibility in the way indicated. Still, the case does go some way toward making philosophical sense of a common but wondrous variety of genuine flexibility: The archer who does not hit the target the first time may simply pull a bit more the next time; the frog who fails to reach a fly circling before it may simply lash out farther when the fly comes round again. Like Tinbergen (see Figure 1), I could be accused of seeing too much similarity between these two protagonists; but at least I would have erred upwardly.

Since the sort of case depicted concerns the exercise of an ontogenically acquired habit, the case's relevance to the issue of instinctivity may seem moot. But even habits can be innately predetermined; and other "principles of configuration," to use the more general Gestaltist term, may share a habit's teleological scheme, exhibit comparable flexibility, and yet be every bit as much a part of the organism's original equipment as are its other species-specific morphological and physiological endowments. Suffice it to say that the case at hand involves an initial proleptic scheme realization toward the attainment of which the organism exhibits many-one developmental plasticity. This makes the behavior that implements that scheme a promising candidate for the office of instinct. Just how might the aforementioned exercise of flexibility put this candidacy in doubt?

Given the case as depicted, the organism's first failed behavior could be utterly predetermined in direction and extent and, hence, utterly predictable (in principle). Since this first failed behavior is, obviously, not well coordinated with its environment, some might charge that the behavior lacks survival value and, so, is not entitled to be called an instinct. But this charge puts too high a premium on success: All that Mother Nature demands is that the behavior should have a passable track record among members of the species. If it does, the behavior type might still qualify as instinctive.

But what about the successor of the failed behavior, the successfully performed variation on the original scheme? Even supposing that this new behavior is predetermined in direction by the mechanism that produces it, one might suggest that the extent of movement in the given direction is not also predetermined, that the "motor" of the movement is turned on and off ad lib by conscious readiness. Of course if this turns out to mean only that the extent of movement is gauged by "sensory input" from distance receptors instead of by preestablished calibrations of the behavioral mechanism, then this lack of predetermination is just relative to one set of factors; and relative to another set of factors, to the input, predetermination still obtains. But another thing that might be meant is that the amount of effort consciousness dictates is not foreordained, that it is not possible even in principle to predict the exact amount, save as a wild guess.

Chapter 9

A Credible Alternative
to Instinct

9.1 Vestiges of Nativism in Flexible Behavioral Tendencies

The true mettle of antagonists to instinct may be their flexibility, but this temper admits of many degrees and varieties, some of which are insufficiently adversarial. A possible case in point is this: An organism might have one scheme in an initial prolepsis and be capable, in given circumstances, of the final proleptic realization of only two alternative schemes. Thus, my frog might have an original scheme for lashing out x centimeters and have the flexibility to do just that or, alternatively, to lash out $x + 1$ centimeters. In some sense, then, this frog might be said to be determined, by its nature, to do one of these two things; and this seems too little opposed to the claims of instinct. The frog's behavior is not uniquely determined but is all too narrowly determined (or circumscribed) even so.

Given that conscious readiness would seem, intuitively, to allow for continuously variable effort, for stretching to any point within a continuous (linear) range of possibilities, the case at hand calls for an explanation of the curious restriction of proleptic outcomes to exactly two schemes. One possible explanation could preserve this intuition by chalking the restriction up to a sort of species-specific morphological "governor," which uniquely translated each distinct effort below a certain threshold into x cm lashing and each effort above that threshold into $x + 1$ cm lashing. Such an explanation would displace but surely not eliminate instinctivity—which would, if the explanation held true, be situated in the genetically ordained governor.

Another problematic case concerns those Gestalist principles of configuration that are purported to be both flexible and instinctive. Quite in keeping with the intuition that consciousness bends nondiscretely (i.e., anywhere along a continuum) to circumstance, the Gestalt-theoretic notion of flexibility effectively avoids any ontic commitment to discrete predetermined mechanical pathways— ones that might, in lieu of flexibility, be supposed to constitute, collectively, particular principles of configuration. Now the Gestaltists' notion of flexibility can easily create confusion about the true character of their theory, about whether it is vitalistic or mechanistic; but the latter alternative is closer to the truth: Gestalt theory posits biologically determined, *deterministic* dynamisms. What the theory calls flexibility is equivalent to what I have preferred to call reactive (indeed, many-one) plasticity: Any apparent bending of preexistent schemes to circumstance is, in Gestalt-theoretic reality, a uniquely predetermined response to a particular set of circumstances and, so, is philosophically equivalent to the selection of one already existing (i.e., initially proleptically realized) scheme among many.

It is not surprising, then, that Gestaltists are quite comfortable—and rightly so—with the idea of a "flexible" instinct. They are misleading, if not exactly wrong, though, insofar as they seem to suggest that native principles of configuration exhibit an initial indefiniteness or indeterminacy. This could reasonably mean that these principles are not precisely calibrated with the environment in which they first unfold (and that only later, in consequence of that unfolding, do they adjust in the direction of greater aptness). But however crudely these principles fit their environment, there is nothing indefinite in theory about just how they will cause the organism to react to that environment.

Far from introducing a note of indeterminacy into the principles of configuration to which they are attributed, such typically Gestaltist features as a tendency to grasp—mentally or physically—figures against a background would seem to suggest some further native ordering of those principles. Similarly, even a genuinely flexible counterpart to a principle of configuration would seem less at odds with nativism were the counterpart to incorporate such features. So, in the hope of enlisting the most undivided opposition against instinctivity, let us look for recruits least possessed of such features.

9.2 Minimizing Instinctivity without Destroying Flexibility

Our search for behavioral tendencies to contest the dominion of instincts might profitably proceed by envisioning a genuinely flexible variant of a putatively innate principle of configuration, and then expurgating it of as many instinctive elements as possible. Excising some of the aforementioned Gestalt-theoretic features is not obviously a threat to the well-being of such a principle, since in place of them (what might be called those native interests), we might posit— as suggested much earlier, in section 2.13—some far more diffuse tendency to find things interesting. But if we cut away too much, we might well be left with nothing even approximating the inherently teleological character of instincts, and would in consequence have found no suitably diametric opposition to them.

Now Lorenz contends that instinctivity is principally a matter of innate environmental information associated with particular native behavioral tendencies, so the most promising anti-instinctive recruits would seem to be those principles of configuration that start out being quite ignorant of their environments. But if that information is contained within the teleological schemes proleptically realized when the principles are set to function on cue, then ridding the principles of that information might do untold harm to the schemes, rendering the performance of them environmentally unsuitable. Indeed, one might even fear that environmentally ignorant schemes for overt behaviors would fail to qualify as teleological at all—which would, in turn, eliminate the principle (or behavioral tendency) with that scheme from serious consideration as a contender against instinct.

The question, then, is, How little environmental information is necessary for a genuinely teleological scheme of overt behavior? Surprisingly little, I suggest: Suppose that the original scheme of action inherent in a principle of configuration is nothing more than an indefinite plan for prehension. Such a scheme is, I submit, sufficiently goal-directed to qualify as teleological, even though that scheme, by hypothesis, involves no environmental information about the nature of that which might be apprehended.

Worrying less about the details of such a scheme than about the basic character of the behavioral tendency that realizes that scheme, one might describe this tendency as a somewhat diffuse, prehensile principle of configuration. Still more broadly speaking, one might further suggest that the most minimal original order that could be supposed to be exhibited in the behavior of flexibly intelligent animals is a kind of acquisitiveness–not to be confused with any (perhaps later to be developed) materialistic character traits. The story of action might then, should one choose to spin a yarn with the fewest (perhaps no) instincts as protagonists, be said to begin with impulsive, prehensile characters that are goal-directed simply in their wont to obtain something, anything at all that might be grasped. One ordinary type of such behavior involves grasping, in the common sense of physically seizing hold of—using one's hand as a sort of grapnel, with fingers as clamps; but the specific varieties of grasping we animals are prepared to do is largely dependent upon our original morphological equipment (and its developmental readiness for use). Thus, the chick grasps by pecking, with its beak as grapnel; the frog grasps by lashing, with a sticky tongue instead of something with hooks or clamps; the duck dabbles with its bill; and so on.

Although prehensiveness is supposed to be in the very essence of Gestalt principles of configuration, those principles are also differentiated by their deterministically reactive plasticity. Assuming for now that prehensiveness and plasticity are compatible, let us inquire whether prehensiveness is compatible with genuine flexibility, too. Given the functional autonomy of an organism, the flexibility of one of its behavioral tendencies is conclusively established by the novelty of the teleological scheme of a final prolepsis in comparison with the scheme(s) of the corresponding initial prolepsis. So one problem with attempting to establish the flexibility of a putatively diffuse (or indeterminate) prehensile tendency would seem to be its very indeterminacy, its initial lack of a definite form. In place of the convenient abstraction of some definite teleological scheme(s) in an initial prolepsis, there might seem to be nothing in comparison with which the scheme

of the final prolepsis might be adjudged novel. This conundrum seems more striking when one remembers that the stretching (or contracting) that constituted the
exercise of flexibility in the case of a first-order intelligent habit was in the same
direction as that habit's original scheme and was merely beyond (or short of)
the extent of that prior scheme. Since indeterminate prehensile tendencies lack a
definite direction as well as a determinate extent, it is hard to see how the earlier
illustration of flexibility could help to make philosophical sense of the idea that
they, too, are flexible.

We have in this puzzle a reflection of some dialectical moves associated with
Dewey's dynamic breakdown theory of consciousness (which for him is also
requisite for flexibility): Indeterminate impulses are first coordinated (by caretakers, perhaps) with vital ends, whereupon diffuse impulses become definite
habits; and only then, if they break down, may flexibility be exercised. But it goes
against the grain to think that things must first acquire greater definition before
they may be said to be truly flexible. The mistake is analogous to a confusion between epistemological and metaphysical concerns: A baseline must be set before
it can be surpassed, but this does not entail that a performer lacks the capacity
to achieve higher before lower levels of performance. Even if their flexibility
could be assessed only once behavioral tendencies acquired definite teleological
schemes, it would not follow that flexibility was absent prior to this acquisition.

The present puzzle is complicated by the fact that the terms in which it is expressed are all so purely metaphysical, so devoid of direct epistemic significance,
that they could hardly have given rise to some excessively epistemological construal of flexibility—a positivistic construal that confuses our way of telling that
there is such a thing with the thing itself, flexibility. Thus, the schemes (in the
initial and final prolepses) that are used to establish flexibility are not objects of
observation as such. The overt behaviors that are attempts to realize fully the
scheme of a final prolepsis, that is, the intention in the very form to be implemented, are observable; but these behaviors may not always accurately reflect
that scheme or intention, since "there's many a slip" between what we intend
and what we actually manage to accomplish. Alternatively, of course, we might
rely on so-called introspective evidence, which, though not indisputably observational, is surely within the bailiwick of epistemology. But suppose that we do
compare our original intention with our final one and discover some novelty in
the latter. This would still not establish genuine flexibility, unless we could be
confident that our final scheme was not also proleptically realized, without our
being aware of it, at the time of our original, conscious intention. All this is
merely to suggest that the very "novelty" whose correct ascription was said to
prove that flexibility has been exercised is not a straightforwardly epistemically
grounded concept, that the proof is at a (metaphysical) level of reality somewhat
removed from the (epistemological) level of appearances.

So, for what it is worth, novelty does not make for an excessively positivistic criterion of flexibility, but novelty might still be thought to suggest an inadequate, overly pragmatic conception of this capacity, as meaning its conceived effects. The main deficiency of this conception would be, for present purposes, its failure to take account of the flexibility inherent in the functioning of an indeterminate prehensile tendency. But is that conception or criterion of flexibility really thus inadequate? Perhaps indeterminate prehensile schemes are enough of a basis for attributing novelty to their own successors and, hence, attributing flexibility to the behavioral tendencies that realize such schemes.

The proposed mark of flexibility is not nearly so defective as it may have seemed, for however indefinite a baseline the original scheme of a prehensile tendency provides, the definiteness of the final proleptic realization of that scheme's successor will stand in sharp contrast to, and so will be novel in comparison with, that earlier indefiniteness. A glob of plasticine need not take on a definite shape before exhibiting flexibility—to take on that shape is already enough of an exhibition. What we have is the epigenesis of a definite form from comparative formlessness, a formlessness with the requisite flexibility to bring about this form or others—depending on the circumstances, on what there is around to grasp. But these circumstances are more akin to opportunities than fateful contingencies that determine the direction and extent of our efforts, else prehensile tendencies' original schemes would be more determinate (in one sense) than per hypothesis they could be.

9.3 The Image of Consciousness in Action

Dewey depicts original impulses as aimless, yet to be vitally coordinated with their environments. He distinguishes any subsequent, posthabituation impulse from thought but calls them twins—born at the same time, "in every moment of impeded habit." [1] Such later impulses are not so aimless: Having been canalized by habitual tendencies, the newly released impulses preserve the direction of those tendencies but now energize consciousness, thought, instead of behavior. Contra Dewey, I should not wish to distinguish consciousness, specifically conscious readiness, from impulse, nor to regard even original impulses as entirely without aim.

My own somewhat speculative story of action begins with consciousness devoid of knowledge about the environment, with conscious readiness to grasp at something one knows not what. This most primitive intentionality is already an initially proleptically realized scheme for prehension—a preparedness to grasp, albeit an astoundingly ignorant readiness. Originally, our gross morphological characteristics might be said to reflect more knowledge about the environment than does our diffuse awareness of it as a field on which to play out our prehensile tendencies, but this order of things is soon reversed, since conscious readiness acquires an increasingly greater facility for grasping, whereas our morphology tends to be more stable, to favor conservation over progress. There is doubtless much truth in Dewey's contention that know-how "lives in the muscles, not in consciousness." [2] Yet when that know-how is employed in ordinary actions it is functionally inseparable, indistinguishable even in thought from the conscious readiness to perform those actions. So-called motor memories have far less autonomy than habit-oriented philosophical psychologies of a Jamesian-Deweyan sort would seem to suggest.

My differences with these theorists may be largely a matter of differing emphasis, but I am inclined to think that they overestimate the prospects for robotization of agents. I would agree that it is possible in some artificially stabilized environment (a factory's assembly line, perhaps) to function largely unaware, but I am not reassured of the cogency of Dewey's own position on this matter when he urges the unlikelihood of this possibility by observing that "in every waking moment, the complete balance of the organism and its environment is constantly interfered with and as constantly restored." [3] Lurking in Dewey's observation is his tacit commitment to the dynamic breakdown theory, and that theory distorts our understanding of the presence of consciousness in ordinary action. Picture a man asleep yet still functioning actively in mundane pursuits: Moment to moment his equilibrium is upset; he wakes up, adjusts his habitual scheme, and then goes back to sleep, having once again turned behavior over to the custody of habit. That this is said to happen in an exceptionally brief span of time does not reassure me of the need to fall asleep in order to wake up again—staying awake, ready to react is not only a more reasonable thing to do, it also appears, phenomenologically speaking, to be just what we are doing when we ordinarily act. Waxing skeptical, I suppose I should have to admit that it is possible I might be nodding off and waking up much more frequently than I could ever hope to know; yet I see no reason to concede that doing so is not only possible but necessary.

Ordinary action, with its fairly definite initial teleological scheme, is performed consciously; and the performance is not, so far as we can tell, intermittent, not punctuated by brief but numerous moments of automatistic functioning. We do not suffer our own actions, and we do not even seem to suffer phases of behavior that further those actions. One dark image that conveys a sense of how consciousness attends action is that of a shadow cast by the scheme of the action: As the shadow of a hand might sometimes make contact with an object presently not reached by the hand itself, consciousness might be supposed to be an anticipation of grasping to come, on occasions demanding some stretching beyond the present reach of activity. But is consciousness really such a shadowy figure that it only shows itself on problematic occasions and is conspicuous by its absence midday in the course of routinely successful mundane action? Like a shadow, consciousness is a projection of the form of that doing the casting—namely, of a teleological scheme; so it is only to be expected that in cases where consciousness is not projected beyond the unfolding schemes its presence will not be patent. Still, consciousness might be worth positing even when its form exactly coincides with a scheme being realized. In terms of the analogy, it might be observed that although an object at midday casts no obvious shadow on the ground below, we have only to pick the object off the ground to observe its shadow—which was, we should think, there all along. Speaking more directly of consciousness' attendance at the performance of mundane actions, it might be said that as we ordinarily act, even in the absence of special care, consciousness coincides with schemes as they unfold (or we unfold them); yet, were we to take phenomenological heed of these proceedings, we should discern that we are aware of what we are doing throughout the course of doing it. Our mental grasping that coincides with physical grasping may not stand out in sharp contrast to it, the way more anticipatory grasping, for example, does; but this is just to be expected and can really occasion only philosophically extravagant doubts.

Earlier (section 6.5), in attempting to fathom the character of the presupposition of conscious readiness by action, I proposed that the former's presence was, strictly speaking, not required during the latter. I am not retreating from that position, only suggesting that performances of ordinary actions are attended by conscious readiness, which, during the acts, does not have an inveterate tendency to nod off and so does not inherently need all the time to be nudged back to attention before being in a position to react properly to what is going on. Placing our action in the full custody of automatic mechanisms, habits, is an extraordinary procedure, possible only in the most rigidly stable (more likely than not, artificially stabilized) circumstances. Ordinarily we do not thus relinquish control, though we do rely on habits' secretly enabling powers to help us exercise that control, and of course sometimes we are distracted (like the proverbially absentminded professor) or are attending more to other phases of our action (anticipating or regretting them, perhaps) than to those phases presently in process of implementation.

9.4 The Role Played by Conscious Readiness

A major drawback of the shadow trope (for consciousness) is its suggestion of idleness on the part of a shade, as if consciousness were a functionally inconsequential ghost in an organic machine. And talk of conscious readiness' being in attendance at ordinary acts does little to make it seem any more efficacious—consciousness of our acts, the shade, has been transformed in this telling into what might seem a passive spectator, one of the members of the audience at the performance of a play. But conscious readiness is no mere spectator, not even one with privileged access to a seat behind the curtain nearest to the actual performers. Conscious readiness is on stage, able to watch the proceedings from a special vantage point—not as a theatergoer or a supernumerary but as the solo performer of the piece. There is no mere coincidence in ordinary action between the ongoing performance and our consciousness of it, for that consciousness, none other than conscious readiness, is our tendency to perform the teleological scheme of the action. Conscious readiness is the proactively flexible equivalent of reactively plastic principles of configuration. Unlike them, conscious readiness is no purely mechanical player in a pathetic spectacle, a fully automated "performing" art.

Despite appearances, conscious readiness is no mere homunculus, either—where homuncularity would be the multiplication of agents (within the body politic of an individual psychological system) beyond necessity, beyond one. If we ever were tempted to construe this consciousness as a separate functionary, an agent of sorts, we would do well to remember that it is we who are consciously ready, who act accordingly. Metaphoric comparisons between conscious readiness and a musical performer might seem to reinforce the idea that consciousness is being viewed as a veritable homunculus, but a more literal comparison or identification might actually weaken this idea by making more plain that the whole person is the consciously ready performer.

Yet even if homuncularity is not in evidence, it might be suggested that what J. Fodor[4*] calls modularity is: The individual teleological schemes in process of being consciously realized might be viewed as distinct modules of psychological functioning, each with its own special purpose, as predetermined by its specific structure, its initial scheme. One cure for this fractured view of ourselves, as consisting of numerous distinct modules, might be to note the commerce among our schemes—the sharing of meanings or purposes, whether expressly, as in conscious deliberation, or tacitly, in consequence of something akin to "the interpenetration of habits."

Of course this treatment may not cure an associated complaint—namely, a certain "epistemic boundedness" (in Fodor's words)[5*] to which even communicating modules might seem prone, a structural limitation of cognitive processing to a nonexhaustive range of topics and pursuits. But given its inherent flexibility, conscious readiness would seem to know, to have, no such bounds. Even after originally diffuse prehensile tendencies set the organism off in a particular direction, after they acquire a definite scheme, their flexibility enables them to change direction; and special purposes subject to modification, specific structures capable of being bent, are not obviously limited to a given range of topics and pursuits. (By the same token, even a prehensile tendency inclined by nature, originally equipped with a fairly definite scheme, is not bound epistemically by that inclination—not if that tendency is sufficiently flexible.) Our sensory apparatus and other morphological organs of prehension may constrain the range of our cognitions, but it is not obvious that conscious readiness itself restricts that range to anything short of the whole, the range of all possible cognitions. The ability to cognize, which conscious readiness by virtue of its prehensiveness exhibits, would only be surpassed by a further talent for cognition without apprehension; but this is no more possible, say, than swallowing food without engulfing it. In principle, then, conscious readiness is not susceptible to epistemic boundedness, though it might be argued that in fact there are some limits imposed by our finitude, our exhaustible capacity for apprehension. But here perhaps a variant of the last analogy misleads, for it is scarcely necessary to engulf in order to apprehend, and we may get a handle even on the infinite, grasp it, with suitable concepts devised for the purpose. Sheer numbers of topics, items for apprehension, may also overwhelm our finite capacity for cognition, so we may be unable to apprehend everything; but whatever upper epistemic bound might thus be set is not topic specific and does not obviously preclude our apprehending any one of the topics, even the loosely defined topic of "everything."

My preceding denial that conscious readiness is homuncular was based on a somewhat Deweyan outlook, but even a patently dualistic viewpoint, of the Cartesian variety, could license that denial—by insisting that any particular conscious readiness is our inalienable property, a mode of our existence as *res cogitans,* as a thinking thing. Of course the notorious difficulty in viewing consciousness in Cartesian terms is that its causal efficacy becomes puzzling, having to bridge a conceptual gulf between nowhere, the locus of Cartesian consciousness, and someplace, the home of corporeal substance. But my appeal to what Descartes took to be inseparable from us does not commit me to endorse his ideas about either the distinguishing characteristics of that which is (I agree) inseparable or the separability of body from mind.

Dewey's antidualistic proclivities might seem to make his views a less puzzling point of departure, though his view of consciousness simply evades the familiar difficulty of Descartes's view: Dewey's dynamic-breakdown theory suggests that consciousness is not causally efficacious in bringing about behaviors— that it is only useful in the planning stage (which might be supposed to require only intramental causal connections), not in the subsequent implementation of the schemes. Still, it might be instructive to note wherein my proposal about conscious readiness' role in bringing action forth most closely parallels Dewey's viewpoint. Impulse, the motive force of a Deweyan habit, is comparably efficacious but importantly different from conscious readiness: Impulse is constrained by habits (or habitual schemes), rendered blind by them until they break—setting impulse free, removing its blindfold; whereas conscious readiness is relatively free all along and only blinkered, not blinded by its proprietary schemes. There is a distinction here between external constraints that must be removed by outside forces, by external circumstances, prior to free movement and internal constraints that can be stretched, somewhat independently of circumstances, by free (though not effortless) movement; and therein lies the essential difference between the Deweyan notion of an impulse ensconced in a habit and my notion of conscious readiness informed by its own habitual scheme. Dewey's account of consciousness renders it as no less free than conscious readiness but not as fully engaged in bringing about actual behavior, and his account of impulse renders it as much less free (than conscious readiness) whenever it is engaged in implementation per se.

Perhaps the least difference between Deweyan posits and my variants of them separates his original impulses from my originally uninformed prehensile tendencies. The difference is that the impulses are said to be aimless whereas the prehensile tendencies are supposed to be inherently (though ignorantly) goal-oriented, to exhibit a diffuse *conscious* readiness to grasp. Now if the point of both positings were to minimize instinctivity, Dewey's theorizing might be thought to succeed better than mine, since (original) prehensile tendencies seem more fraught with innate content than do aimlessly impulsive ones. Of course he has also to account for the subsequent appearance (in the course of cognitive development) of intentional consciousness, but that is where his dynamic-breakdown theory comes to the rescue—explaining such consciousness as emergent under specified conditions. To its credit, the theory does plausibly suggest that the directedness of what emerges might be a sort of persistence of the directedness of the habits that break down, but, on the debit side, the theory's appeal to emergence seems all too much like a suggestion that their being pulled out of a hat explains where rabbits come from.

9.5 Consciousness in Development

To make more sense of cognitive development, to avoid the mystery of emergence, I posit sufficient consciousness at the beginning of the developmental process to preclude any need for later emergence; though, for all I know, cognitive development does not make this sort of sense—any more than many other natural causal processes do. Still, the drama of development is philosophically (if not scientifically) more rewarding when consciousness is no mysterious deus ex machina but a principal performer throughout the proceedings.

Suppose we take a tack somewhere between Dewey's view of how consciousness develops and the view I am defending: Consciousness might be seen as initially formless, lacking even a diffuse prehensiveness, but progressively intentional, borrowing its earliest form from the habits set up for us by our caretakers. This alternative would avoid the mystery of emergence and does hold out some promise of diminishing the commitment to innateness, but the promise may remain unfulfilled: In place of original prehensiveness, this alternative is freighted with a potential for the acquisition of directionality, an innate potential that is not obviously a lesser commitment. Besides, it is not obvious that something lacking directionality even qualifies as consciousness, so perhaps the would-be alternative view is just Dewey's view in disguise. Finally, alternative or not, it is also less than obvious that subsequent "consciousness" with mere directionality could ever aspire (save via emergence) to prehensiveness, without which later cognitive developments (such as truly intelligent functioning) would be impossible.

Despite its faultily fatalistic view of flexibility, Gestalt theory does stress prehensiveness as essentially characteristic of every principle of configuration. Nevertheless, I should like further to contradistinguish conscious readiness from a principle of configuration, on the somewhat surprising grounds that such a principle, as depicted by the Gestaltists, would still be insufficiently prehensile to account for genuinely intelligent functioning. Notwithstanding the prominence they give to it, Gestaltists fail to assign grasping its proper role in problem solving: Grasping a problem is cast as little more than the realization of a solution. The conceptual space between understanding a problem and finding its solution is filled by a dynamic tendency toward closure, and that tendency is a solution in the making, along predetermined lines.

There is some plausibility to this view of problem solving, for we do often (perhaps always) only fully appreciate problems as we discover their solutions. Nevertheless, it would be a mistake to deny the major role played, in all cases of intelligent problem solving, by a prior (albeit less fully appreciative) grasp of problems without solutions. Gestalt theory effectively mimics problem solving by substituting dynamic plasticity for genuinely intelligent flexibility. The point, then, of marking off problem grasping as a distinct phase of the proceedings is lost: teleological schemes will bend automatically, with or without attendant awareness of the need to bend them. Why should the Gestaltists pay any attention to the "moment" between problem grasping and problem solving? On their view, the organism's future is all mapped out for it, and the smoothly efficient functioning of the dynamisms, whether it is prolonged or instantaneous, whether conscious or unconscious, is all of a piece. This functioning is, I submit, a piece fashioned of whole cloth.

Cases of missing goal-objects show, indirectly, part of what is wrong with the story told by the Gestaltists: How can a dynamism stretch toward the unknown or the unapprehended? This question soon arises when attention turns to the details of cognitive development, which might be characterized, in basic accord with Gestalt theory, as a matter of prehensile tendencies' becoming increasingly adept at obtaining their goal-objects: First, things in contact may be grasped; next, things close to hand; then, things far away; and, still later, things not there at all. Be this course of development as it may, it appears to involve a special leap forward when it comes to this last step, so remarkable in comparison with its precedents. But this leap, wondrous as it is, is taken far sooner than the Gestaltists seem to think: The organism is *not* directed to the goal-object near to hand by a determinate dynamic tendency for behavior to spread out in the direction of a stimulus afforded by the presence of that object. Rather, the organism's indeterminate prehensile tendency directs it, however ineptly, toward a goal every bit as absent as a missing goal-object. Consciousness grasps the object as a goal yet to be attained, even should the organism's "contact receptors" already be in contact with the object; for it is the further grasping of that object, not the object itself, that is the organism's goal. All the purposive behaviors mentioned in the above sketch of development, not just the behaviors of the last stage, are guided by the grasping of not-yet-present goals by conscious readiness.

Accordingly, given some memories of past, once-present goal-objects to latch on to, the worldly-wise leap toward absent goal-objects becomes almost as commonplace as the more modest jumps that may, cognitively-developmentally, precede it. All the jumps remain wondrous, but the last one loses its especially puzzling appearance. Looking still farther along the course of development, we might now, for similar reasons, find less puzzling the grasping of a problem with an unknown solution. The puzzlement starts with the thought that we must, to grasp the problem aright, get enough of a hold on its unknown solution to be guided by it in the intelligently flexible pursuit of it; but what sort of hold is possible when, as memory correctly serves, we never did behold this thing (the solution) as the object of such a desire (to resolve the problem)? The Gestaltist answers, as before, that insofar as we are capable of a resolution, some predetermined principle of configuration is already possessed, proleptically, of the problem's solution. This answer does allow for learning, experiencing, and other ontogenetic processes to intervene in what leads up to the resolution but still suggests, too fatalistically, that the organism has completely grasped what will be its solution, at a moment when the whole point was for that organism to grab hold of the problem in the absence of a determinate solution.

Some of the puzzlement is alleviated by noting that every solution to a problem, however routine, is unfully apprehended until the moment of resolution; but of course other, less puzzling cases do involve something (e.g., a recollected goal-object) to latch onto in place of the goal/solution. In the case at hand, all we do have is a problematic consciousness focused by an incomplete teleological scheme, a searchlight not yet illuminating anything more definite than clouds but casting about in various directions—a beam on a cloudy night, but a directed beam at that. Yet how would a mere beam *know,* even if it illumined something, that the search was a success? The question shows the need for a more suitable image, perhaps this: The apposite problematic consciousness is a specialized organ of apprehension, an organ whose chief purpose it is to grasp a definite solution and whose specialization consists in its delimited capability thus to apprehend, to be apprised of a solution. How does this quasi-figuratively depicted consciousness know success, know that a solution is in hand? Quite simply by way of apprehending that solution—though further means of making sure (say, by direct examination of the fit between the problem and its solution) are also available to some organisms, possibly to all of those capable of advanced problem solving in the first place.

For all its sophistication, the grasping of an unsolved problem shares with the most naive conscious readiness a certain indeterminacy with respect to goals. Of course the former, developmentally later, mode of apprehension is more informedly specialized than the latter; but even when the former is so specialized that it can truly apprehend only a single goal, the unique solution to its problem, there is no guarantee that it will. This still problematic consciousness has grasped the incompletion of its proprietary scheme, grasped this as a felt need for apt closure, and is ready to take some of whatever steps it can to satisfy that desire; but it, this conscious readiness, has not yet (even proleptically) attained its ultimate prehensile purpose, which is still, for consciousness, a somewhat indeterminate goal. There is a special, unintended aptness to talk of *grappling with* a problem—as if to say that the incomplete teleological scheme that is the intentional content of the problem is a device used in trying to grab hold of a solution. Conscious effort, a phase of conscious readiness itself, may take its marching orders from its specialized scheme but is able to stretch them, to bend the rules, attempting thereby to achieve a resolution. A developmentally senior effort of deliberative consciousness might go so far as to borrow parts of other schemes, using these parts to supplement its own scheme, perhaps even to complete it.

9.6 Consciousness as Active Prehensiveness

My view of the consciousness implicated in our actions has it that a certain prehensile tendency is the very form of that consciousness. But another opinion, the intentionality view, holds that all consciousness is just a sort of directedness toward possibly nonexistent objects, an awareness of them. My disagreement with this view needs spelling out. I do find paradoxical the suggestion that the directedness must have its object even if it happens to be a nonexistent one, and I take my alternative to avoid the absurdity. Indeed, the last section tried to show how my prehensility view of conscious readiness might avoid a special, advanced case of such absurdity, part of what is known in philosophic circles as Plato's paradox of inquiry: How can you search for what you do not know, since you do not know what you are looking for and, so, could not even *recognize* it if you stumbled upon it? My half-figurative reply was that you could be well equipped, well prepared with a specialized grapnel, to reach for and grasp the thing for the first time, and that your specialized cognitive equipment might simply make the thing found feel right to the grip, seem (because it is) eminently, aptly graspable.

Curiously, *direction* would seem to imply less need for an object than *prehension;* but *direction toward* is the (object-implying) concept employed in the intentionality view, and my view might make do with the concept of *grasping for,* which, as a drowning man might know too well, is applicable in the absence of an object. Still, my view occasionally requires something ontologically iffy to hang on to, too—as when intelligent problem solving is expressly undertaken—and affords on such occasions (functionalistically abstract, variously realized) objects, incomplete teleological schemes, suitable for the purpose.

Another area of disagreement is that the intentionality view seems too compatible with a passive spectator conception of consciousness. Conscious readiness in particular should never be so conceived; it is ever an actor, never simply a passive spectator. And, as if to oppose the intentionality view most fully, I should like to propose the same of consciousness in general: All consciousness *acts* in virtue of its essentially prehensile nature. The opposing view sometimes verbally suggests a not-so-passive conception, by talking in terms of intentional acts of consciousness; but the view's heart is not in this way of speaking.

The passivity at the core of the intentionality view is its compatibility with a purely reactive sense of directedness—which is not to say that the view is essentially wedded to such a sense. Mechanistic teleological plasticity, as presented in sections 5.3 and 5.4, bears witness to this compatibility by satisfying the conditions for directedness toward (even nonexistent) goals and goal-objects while involving nothing more than mechanistic reactions to circumstances. A terminological rebuke is now in order: The term *intentional(ity)* has wrongly been invoked by the mere-directedness-toward view, since there is nothing distinctively intentional about the operations of a system that exhibits mechanistic teleological plasticity. In the realm of behavior, the term *intentional* should be reserved for directedness in consequence of conscious readiness in furtherance of a teleological scheme; nothing else squares with *what it means* to speak of action as opposed to mere (albeit teleological) behavior.

To extend my observations to realms other than that of overt behavior, I might further note that, on the (so-called) intentionality view, visual perception, for one, could be understood in exclusively passive (or reactive) terms: There is no incompatibility between, say, a mechanistically causal theory of perception and the view's suggestion that, insofar as seeing is a consciousness of something, seeing is a mere directedness toward that something. After all, such directedness might consist of an Automaton's teleological plasticity with respect to getting or sustaining a visual fix on the object, where having such a fix might be defined in terms of certain stimulations of the optic nerves or sensors.

"To our consciousness," Gottlob Frege remarked, "the line of vision from the eye to the object is straight."[6] This might seem to hint at a more properly proactive understanding of vision, as actively directed outward; but however liberally interpreted, Frege's observation cannot be said to go far enough in one fundamental phenomenological respect: Vision is not just *perspectival*, but also what may be called *prospectival*. Vision provides a view (in a particular direction, to be sure) of objects within the "reach" of the eye; and our consciousness of visually perceived objects, our seeing of them, is a successful prehension, a visual apprehension over and above mere sensory directedness toward or even directedness upon them. Therein lies vision's essentially proactive character.

But can there not be purely reactive grasping—say, by a robotic hand, closing in response to stimulation (by an object) of electronic sensors? Not really, I am inclined to say, for all the temptation there might be to extend certain conceptual courtesies to devices shaped a bit like people: There is no effort, no real attempt that could qualify as grasping for, no real accomplishment to stand as grasping of; there are only some cold metallic fingers encircling an object.

Inasmuch as consciousness of something is (in Ryle's term) an *achievement*[7] or (in F. N. Sibley's philosophically divergent term) a *retention*,[8] consciousness presupposes some *doing* with respect to which the achievement was the upshot or the retention was the continuing consequence. To illustrate the use of these terms, it might be said that sudden noticing of is an achievement of consciousness, while protracted awareness of is a retention. In either case, the doing is nothing less than the very act of consciousness—a grasping or holding (i.e., keeping in grasp), respectively. But consciousness need not be altogether an achievement or a retention; consciousness may also be an effort to grasp or a grappling still unsuccessfully with—in which cases, the doings of consciousness are even more obvious, while its prehensiveness is no less so.

All modes of consciousness are doings wherein it exhibits its prehensile nature. All exhibitions via proactive doings of prehensiveness are modes of consciousness. Purely reactive grasping is so called merely by common courtesy—though most plausibly so in cases where there is nonintentional teleological plasticity with respect to the putative prehension. It may be stretching a point to insist that such cases involve no doings, since, for example, Venus's-flytrap is doing something when its leaf-lobes close around a fly that touches their hairs; but the less exaggerated point is just that any such doing is equivalent to a suffering, an undergoing, not an undertaking. The doings of the plant do not seem of a sort to qualify the plant's engulfing of a fly as active grasping of it, and merely passive prehension, should we choose to call some reactive doings that, is not an act of consciousness.

Yet are not some of what we are pleased to call our acts of consciousness fairly passive occurrences? Consider sorts of cases just used to illustrate retentions by and achievements of consciousness: We sometimes say—in an odd reversal of the averred direction of apprehension—that an especially fascinating object that has our protracted attention is holding our attention; and although active searching may (sometimes) precede our sudden noticing, the noticing itself always seems something of a spontaneous occurrence, an almost automatic apprehension. But the apparent passivity might be something of an illusion based on an implicit contrast with full-fledged actions, by comparison with which the above cases are bound to seem relatively passive. Indeed, the cases should not even be described as intentional, for though each involves a prehension guided by a teleological (or, at any rate, prehensile) scheme, that scheme is not so much implemented by means of consciousness as realized directly by it; the cases involve no conscious readiness *in furtherance of* their prehensile schemes. The acts of consciousness in question are (as Blake calls Adam) a "limit of contraction"; their activeness is no more than their specific apprehensions of their objects.

9.7 Native Conscious Readiness: A Concession to Leibniz?

However diffuse the consciousness implicated in our earliest actions, it (like all other consciousness) is proactively prehensile; and this might seem a sizable concession to instinct. For though this may displace other, more instinct-oriented conceptions of overt teleological behaviors in favor of an intelligence-based understanding of them, this understanding must admit the innateness of the intellectual powers, the originally diffuse prehensile tendencies, themselves. But despite the rationalistic (or nativistic) appearance of this concession, it is not a clear victory for instinctivity.

Following out the evident parallel in Leibniz's philosophizing, we should note his insistence that "the mind [or innate intellect] at every moment . . . already thinks confusedly of all of what it will ever think distinctly." [9] This remark might be interpreted to say that all of an intellect's future ideas—including all its intentions—are proleptically realized at the very start of its existence. But since he allows for some measure of difference between the (confused) initial and (distinct) final state of these ideas, the remark might be supposed compatible with the assertion of flexibility. Thus, a diffuse prehensile tendency might be an originally confused form of thinking that, as subsequently bent to circumstance, at last becomes a distinct form of thinking.

But Leibniz also insists that mental capacities cannot be wholly (and passively) indeterminate, for then they would be unequal to some of the very tasks they perform—namely, cognizing necessary truths in the face of merely contingent (truths of) experience. Since induction from particular experiences could not yield necessary, universal truths, he reasons, "the mind has a disposition (as much active as passive) to draw them itself from its depths; although the senses are necessary in order to give it the occasion and attention for this, and to carry it to some rather than to others." [10] This "disposition [is] an aptitude, a preformation, which determines our soul and which brings it about that [innate necessary truths] may be derived from it." [11] Perhaps one could not adapt his principal reasoning (about necessary truths) to the case of intentions, but note the subsidiary connection educed between preformation and determination. When he speaks of preformation's *determining* the mind he must, to avoid triviality, mean more than *defining;* to avoid irrelevance to the topic under discussion, mean something other than *setting limits to;* and, to stop short of fatalism, which he does not here espouse, mean less than *setting one irrevocable course for.* What he means is that preformation gives a particular aim or direction to the mind—a direction not unrelated to what issues forth, to how it issues. Although Leibniz is not patently fatalistic about whether a particular item will be derived, he seems convinced that whether any particular item could be derived is determined by what is already prefigured in the intellect. This is nicely compared with logical deduction from a set of (perhaps unexpressed) premises, but such a comparison makes it difficult to sustain the flexibility-compatible interpretation proposed above: The exercise of flexibility would entail a genuine novelty in what we come to think distinctly (in contrast to what we earlier thought confusedly), yet novel truths are not forthcoming from something akin to deductive inference.

Or are they? The deductive principle of addition, which licenses the inference of "p or q" from "p," might seem to afford a counterexample; but one could rejoin that whatever truth is apportioned to "p or q" in logical consequence of "p" is just the truth of "p," not some novel truth. Still, what of the allegedly valid inference of "p or not p" from "q"? This apparent counterinstance does not lend itself to any like rejoinder, though one might otherwise reply that it is misleading to claim the case is one of inference per se, as if to claim that the truth of the conclusion has meaningfully been guaranteed by the premise. One might as well say—along with devotees of what is called (with no intended irony for Leibniz) "natural deduction"—that a logical truth such as "p or not p" is deducible from no premises whatever; and this seems far from helpful to Leibniz's position, since logical truths are necessary ones that ought, by his light of reason, to be derived from preformed inner resources of the intellect. The likely Leibnizian reply is that some such resources, innate principles of inference, should still be needed to achieve the feat of deducing logical truths from naught. And there is some undeniable merit to this reply, understood as saying that insofar as we can deduce logical truths from inner resources, these resources must be preformed either as (perhaps "confusedly" thought, i.e., unexpressed) premises or as (perhaps only tacitly known) principles of inference.

But deduction may not be the only way to derive ideas from other ideas (or intentions from other intentions), and should there be a less determinate means of doing so, the case for preformation would collapse. This brief excursus into Leibnizian dialectics may seem relevant only to intentions that happen to be necessary truths; and, with the debatable exception of some instantiation of the Categorical Imperative, there may well be no such intentions. But Plato's paradox of inquiry, the philosophic engine of Leibniz's rationalism, extends his case for preformation to all ideas, contingent as well as necessary, and, so, to all intentions. Accordingly, the lesson I draw from Leibniz pertains to the issue of instinctivity, understood here as a controversy about the innateness of intentions. The lesson is that a sure way to defeat the innateness (or preformedness) hypothesis is to deny the determinateness of the ontogenetic processes that give rise to the structures alleged to be innate.

Deductive inference, viewed psychologistically, is one sort of determinate process that can be used to argue in favor of the innateness of items arising from it; Noam Chomsky's full complement of "rules" that transform the very deepest "deep structure" into the "surface structures" of language constitute another sort—one that, given the enormous rule-wrought difference between the types of structure(s), might best be said to argue for the *just derivatively* innate character of the latter type, the surface structures.[12*] Further argumentation is, of course, also needed to preclude the possibility of external origins for the original, deep structure—just as Leibniz found it necessary to preclude the possibility of acquiring necessary truths "from experiences or from the observations of the senses."[13] The best (non-Leibnizian) strategy for precluding such possibilities would be to demonstrate the many-one plasticity of the developmental processes giving rise to the originals whence are derived, determinately, the end products.

If this fuller argumentative strategy could be deployed, there would be an undeniable residuum of innateness in those end products—even if there were also another variety of indeterminateness coexisting with the rule-accordant determinateness appealed to in the argument. Thus, even if there were something absolutely chancy about when an organism might "follow" its determinate rules, if it did do so, then the results would be innate outcomes—coming from the organism's own nature (coming from there even though first taking a detour, so to say, through environmental variables, such as samples of a natural language, that were partially determinative of the rules themselves). This point about ultimate origins within the organism is particularly evident in cases where the outcomes (e.g., logical truths) are clearly like the endogenous originals (e.g., the deductive rules) whence they are derived—where the offspring show their parentage on their faces; but the point holds true even when the "likeness" is detectable only by tracing the orderly, rule-according, generation/transformation from the original—when we know *how*, by what determinate steps, the changes have been wrought. The moral, in both sorts of cases, is not that indeterminateness does not preclude preformedness, only that the indeterminateness may have to dispatch other determinateness in order to do the precluding.

Returning to the case of originally diffuse, proactively prehensile principles, it may be argued that inasmuch as they are genuinely flexible—indeed, so essentially so that they cannot be (overtly) fully realized without exercising their flexibility—these principles afford no basis for the determinate derivation from them of distinct successors. So, even if we are forced to concede the innateness of these principles, we need not concede the innateness of their (novel) successors—the parentage of these successors may be traceable, but they are far too sportively ad lib for any of them to qualify as a uniquely prefigured outcome of primitive conscious readiness.

Nonetheless, one commonality remains: The otherwise novel outcomes (or articulations) are as prehensile as they were in their former guise. So, there is at least one innate ingredient in the otherwise novel outcomes. But since this particular ingredient, the active prehensiveness implicated in our actions, is requisite for flexibility, the fount of all anti-instinctive novelty, admitting the innateness of that ingredient is only the most minimal—least still philosophically comprehensible—concession to instinct. A thoroughly diffuse prehensile tendency is the most diametrically opposed, dialectically antithetical posit available to combat the pretensions of instinctivity. As it develops, moreover, this prehensiveness, as conscious readiness in the service of sophisticated teleological schemes, is also indispensable to what we own to be actions.

Actions, however, might yet prove to be more instinctive than we know. Suppose that our original active prehensiveness is not so diffuse, that it has a native bias toward certain types of prehension—say, the apprehension of an object moving across a stationary background. Insofar as this bias were to preclude contrasting types of prehension (say, of a stationary background), this bias would clearly be another innate component of action. But what if the bias were not so pronounced? What if extra effort could bend the biased scheme every which way, à rebours? Less diligent actions in accord with the bias might still qualify as instinctive, but nature would have lost absolute control of them.

To reprise what is needed to oppose the rule of instinct, let us return to Leibniz's suggestion, in effect, that preformation implies a direction of the mind such that its innate contents could be derived from it. If the direction of an organism's "mind" is set by a mechanistic teleological plasticity, then those schemes that transpire as the plastic process unfolds may be read back into the organism, be understood to have been proleptically realized from the start; so, this direction is such that innate teleological schemes could be derived. Yet even some nonmechanistic directions of a "mind" could be sufficiently determinate with respect to content also to allow for such reading back—for example, a direction involving chancy but always deductively valid inference; so mere indeterminacy may not be able to overthrow instinct. What is needed is not just indeterminacy but also novelty—in a word, flexibility. Genuine flexibility empowers an organism to act in a fashion, an organism's mind to veer off in a direction, such that innate contents could not be derived.

A problem for this conclusion is posed by Chomsky's claim that "a person who knows a specific language has control of a grammar that generates (that is, characterizes) the infinite set of potential deep structures, maps them onto associated surface structures, and determines the semantic and phonetic interpretations of these abstract objects." [14] Chomsky stresses the great novelty of these deep structures and their associated surface structures, and this novelty might be thought equivalent to that which is requisite for flexibility: No infinite set may be proleptically realized, it might be argued, since such a realization presupposes the incoherent notion of an actual infinity of objects. [15*] But despite what might seem the undeniable flexibility of normal language users, the very (recursive) rules they have at their disposal, to generate endlessly novel but appropriate linguistic structures,[16]* appear determinate enough to argue for the derivative innateness of these structures. Were these rules to operate finitely deterministically, uniquely determining specific linguistic responses to each of a large but finite range of circumstances, the problem for my conclusion would not arise: Each of those responses would have been proleptically realized after all, so there would be no problematic novelty combined with innateness. But if, as Chomsky's attribution to the language user of "control of a grammar" allows, the person is a Performer rather than an Automaton, then notwithstanding their apparent novelty, the linguistic expressions might be credited with substantial innateness.

My solution, however, is to deny their novelty—by suggesting that these expressions' underlying deep structures are proleptically realized as a finite device programmed with recursive rules, a "[system] of rules with infinite generative capacity." [17] "Novelties" thus prefigured, already writ in rules that, in turn, are embodied in a (neurophysiological) device, are not so novel as all that. The Performer equipped with and ever constrained to employ such a device could still exhibit genuine flexibility—of course not with respect to the character of the deep or surface structure of expressions, but apropos of the occasions on which these expressions are actually used. Thus, intelligently devised uses of old (or at any rate prefigured) structures would be novel performances, exercises of anti-instinctive flexibility.

Advocates of instinct (more generally, of biological determinism) might now be prepared to agree that flexibility is what's needed to defeat them,[18]* but they might also begin to doubt its ability, given its family ties to causal indeterminacy, to withstand the rigors of philosophical-methodological battle.

Chapter 10

Beyond Nature-Nurture, But Nature Above All

10.1 Depolarization of Fate and Freedom

Purposive behaviors presuppose conscious readiness, and in all but the most extraordinary, mechanically habitual ones it is actively involved. All ordinary purposive behaviors are proactively prehensile: Each of their goals, each of the ends-in-view en route to them are to be attained, reached, grasped, by way of a teleological scheme—best understood as a *modus prehendendi* in action or, sometimes, only in thought. Where there is conscious readiness, there is flexibility. Where there is flexibility, no behavior pattern is entirely automatic and pre-ordained—none is inescapably instinctive. In praise of idleness, I might remark its conceptual tie to intelligent capacities: That which empossibles the flexibly adaptive pursuit of goals at once empossibles not pursuing them. But advocates of the Protestant work ethic might be cheered, too, that this same conscious readiness also empossibles some greater effort than fate would allow. This consciousness, Nature's apparent device for the guidance of some behaviors toward their organism's vital goals, gives organisms far more leeway than might be bargained for on the hypothesis that the device is simply a product of evolutionary mechanisms. Any such leeway would seem to go against not just instinct but Nature itself. Can this be?

Possible or not, the position is not nearly so implausible as some other views that elicit the same query. Thus, unbridled voluntarism, which raises free will to the status of an undiminishable first principle of action, sees consciousness as a pure Composer free to play any tune of its own devising. But this philosophy willfully forgets that the world is ready-made, and our tunes, even if we could create them out of nothing, would not always be well played, even playable in this world. Our fate is partly writ by circumstances beyond our control: Some tunes may not be heard in the din of history or the poor acoustics of our habitat; some may fall on deaf ears; some may not appear as pleasant to our ears as to our imaginations. We might lack the requisite talent or instrument to achieve the desired effect of our self-bidding of behavior. Then, too, even if fate otherwise permits, we might not put enough effort into the practice needed to achieve a satisfactory performance of our compositions.

To bring forth adept behavior, consciousness needs more than a teleological scheme, be it largely self-devised or not. Consciousness must first learn some lessons from the world before hoping to cope with it effectively. These lessons need not lead to beliefs (or other so-called propositional attitudes) concerning the world but must afford a measure of know-how, some developed mastery without which consciousness is not a profitably efficacious tendency toward overt behavior.

A major blunder of Jean-Paul Sartre's special voluntarism, his existentialism, is to have denied the reality of habits ensconced in consciousness. His cagey nondescription of it as transparent may rightly remind us that consciousness is something seen through—that apprehension takes place by way of it—and that it does not lend itself to self-apprehension. But this transparency should not blind us to the support provided by our habits *in* consciousness. It is one thing for a tightrope walker not to look for the transparent rope beneath his or her feet, but it is quite another for the performer to deny its existence, to believe him- or herself able to negotiate the abyss without a rope. If, changing figures once more, we say that consciousness is a clear stream, we should duly note the presence in that stream of powerful currents and eddies along certain lines—directions set by full-fledged teleological schemes or, at a minimum, by some originally rather diffuse, abstractly functional proprietary structures of consciousness.

Human (and possibly some brutes') existence cannot be said to precede the proactively prehensile essence of consciousness; and, if such existence is understood as viable, relatively autonomous functioning, then it cannot even be said to precede a more sophisticated, teleologically schematic and habituated character of consciousness. Ironically, given his atheism, Sartre's image of humanity is all too godlike: It is bad faith, he says, for the likes of us to suppose our freedom constrained by anything more than choices that we freely make and remain free to revise. Later developments in his thinking allow that harsh economic realities, certain social and material environments, make human freedom no more than an illusion. This goes too far yet not far enough: We are constrained by our environments, but they are also indispensable to effective action and almost always afford opportunities, however limited, for variable effort to play a part; we are also, he neglects to say, internally constrained by habits, environmentally efficacious tendencies to act.

Of course this neopragmatic positing of habits in consciousness might be supposed, ultimately, to deprive it of its flexibility and so to reintroduce the notion of instinct—as a second nature if not a first. C. S. Peirce conjectured that laws of nature can be regarded as the crystallizing habits of the universe;[1] if so, particular organisms coming to behave habitually might thereby doubly reconcile themselves with Nature—becoming more naturally lawful themselves, learning to cope with Nature's other habits. But although our habits may make our actions less chancy, this need not mean that as we accumulate habits we become automata. Our amounts of effort expenditure, for example, might become as regular as the routinely successful patterns of behavior with which the efforts are coordinated; but for all the active inertia this may create, it is not obvious that flexibility is in consequence destroyed: Greater or lesser amounts of effort might still be possible. Indeed, habits in general may (as Dewey might say of intelligent habits in particular) serve more to "improve our aim"—another sense in which our actions may be made less chancy—than to render them automatistic.

A different, common yet implausible view used to challenge the hegemony of Nature would have us free to depart from Nature's dictates insofar as our actions were the result of free volitions. For this to work, the critics of the view point out, these volitions, our acts of will, would require prior free volitions, else those acts themselves would not be free. This point is well taken: The view quickly leads to an infinite regress to which (pace Leibniz[2]) even the freedom of God would succumb, since at no volition along the way would free will be secured. So, even if this infinitely capable Being had managed to perform the infinite series of prior acts, God would not thereby have gotten anywise toward freedom of the will.

My position requires, for freedom's sake, only that we exercise our flexibility, not that we freely choose to do so (though we might do that, too, should we exercise our flexibility in the context of deliberation). Suppose, for instance, that in working at some task I try harder than I ever have, even though the circumstances have not significantly changed since the last time—say, a moment earlier—I was occupied with that task. How much harder I now try is not fully predetermined; it is, however, determined by my conscious readiness—which may from habit expend a definite amount of effort but which is inherently flexible and so able to vary that amount.

Effort, though, is not the only variable of action flexibly determined by conscious readiness in furtherance of a scheme. J. L. Austin notes (apropos of "the suggestion that . . . 'I could have done X' means 'I should have succeeded . . . if I tried' ") that in a case where I say I *could have* holed a putt I did not sink, "It is not that I should have holed it if I had tried: I did try, and missed."[3] Trying harder, expending more effort, I should also note, may not be at issue, either; for among the things under my conscious control (not, like earth tremors, outside of it) are things that can in principle be varied independently of the amount of my effort: In saying "I could have . . . ," I might mean this with the tacit proviso, say, "had I remembered to vary, to loosen, my grip—as the golf pro advised me to do."

A separate, big question is whether what I mean can possibly be right, whether I (as I was then, in those same circumstances) could have done differently, successfully, what I did do unsuccessfully. Austin observes that "a modern belief in science, in there being an explanation of everything" inclines us to conclude that such a thing is not possible, since "surely there *must* have been *something* that [previously] caused me to fail."[4] More specifically, this is a belief in a ("natural") *deterministic causal* explanation of everything that has a direct bearing on behavioral events, including the failure to sink the putt.

The natural world is much with us, to be sure, and purposive behaviors are a part of it; but does this imply that Fate prevails over Freedom? The world where conscious readiness dwells is a place, an "image," made manifest to me in attempting to comprehend my locus in the natural world. My prehension has been informed by what I know or think I know about this world, behavioral sciences, philosophical viewpoints, and so on. The world in which I apprehend a reasonable image of myself would seem to lie between two other possible but less personally comprehensible worlds, the one of machines and the one of gods. My world, if not best, does meld desirable features of the other two: causal efficacy combined with enough liberty to make some personal use of it.

Instead of adverting so obscurely to separate worlds, I might portray them as polar opposites between which natural agency is possible in this, the only world real to us. A proper understanding of ourselves must steer clear of full Fate and total Freedom, points beyond all human bounds. But there is no need to bridge the gap between, to live uncomfortably suspended betwixt these particular extremes; there are some less antithetical counterparts marking limits internal to the domain of human affairs. Thus depolarized, our real fate is just the vast, inestimable contingency of circumstance (the social and physical environment, our own human and organic frame), and our real freedom is an inexactly estimable but no doubt (inherently as well as circumstantially) limited flexibility.

10.2 Creedal Determinism

My last remarks agree, by design, with much that Emerson says:

> The Circumstance is Nature. Nature is what you may do. There is much you may not do. We have two things,—the circumstance and the life. Once we thought positive power was all. Now we learn that negative power, or circumstance, is half. Nature is the tyrannous circumstance, the thick skull, the sheathed snake, the ponderous, rock-like jaw; necessitated activity; violent direction; the conditions of a tool, like the locomotive, strong enough on its track, but which can do nothing but mischief off of it, or skates, which are wings on the ice but fetters on the ground.[5]

But there is much here that should be disagreed with, too: Emerson's remarks about thick skulls and rock-like jaws are either unilluminating metaphors or damning evidence that his philosophizing is ill informed by the phrenology of his day. And while Emerson may concede to life "the positive power," half the control over the course of animal events, he seems satisfied, ultimately, to regard the course itself as fully necessitated by that power combined with circumstance. Emerson is, I am almost certain, a compatibilist, but Nature does not force this view on him.

Compatibilism is the doctrine that "free will is compatible with determinism and inconceivable without it." Compatibilists, I contend, just have not tried hard enough. Emerson preaches some such doctrine: "Let us build altars to the Beautiful Necessity. If we thought men were free in the sense that in a single exception one fantastical will could prevail over the law of things, it were all one as if a child's hand could pull down the sun. If in the least particular one could derange the order of nature,—who would accept the gift of life?" [6]

This all too genteel sentiment is alien to me. I answer, "I would." My childishness would not extend to pulling down the sun, though the philosophic motivation of this essay is, in part, "to rage against the stars." Emerson's faith is in a "Necessity which rudely educates him to the perception that there are no contingencies; that law rules throughout existence; a Law which is not intelligent but intelligence;—not personal nor impersonal—it disdains words and passes understanding; it dissolves persons; it vivifies nature; yet solicits the pure in heart to draw on all its omnipotence." [7] This Romantic creed, this transcendentalist conceit, grants intelligence no real power, only some hope of recovering a preestablished harmony and the consolation of a false idol, Nature in warped philosophical perspective.

We now look to science with the same sappy gleam in our eye that Emerson, transparently, had in his as he gazed upon Nature. Earlier, spiritual dogmas have been replaced by mechanistic articles of faith. Chief among them is determinism, dogmatically embraced as a proper measure of all things above a certain subatomic size. It seems a questionably visionary yardstick. Evolutionary mechanisms and Galilean principles before them are more substantial devices, potent destroyers of creaky old purposive explanations for too much under the sun. But while it is scientific progress to deny purposes of strictly biological and natural physical phenomena, it is just some new creed that rejects the purposiveness of all behaviors of all animals.

The outcome of essentially deterministic evolutionary processes has been a bountiful harvest, but there is no good reason to aver that as Nature sows, so it reaps, that evolutionary mechanisms must surely give rise to behavior-producing mechanisms. Only a benighted view of causality would insist on projecting into the effect, the harvest, so direct a reflection of its cause. Why be so philosophically prescriptive, prior to a careful inventory, about the nature of the items in this fairly random harvest? If among the fruits of natural selection is consciousness, we should not be quick to deny its special efficacy and possibly attendant advantages for survival.

It is possible that numerous species-specific teleological behaviors have evolved and that in any case where consciousness figures, it is a superfluous feature of the mechanism for the selection and implementation of the behavioral scheme—that automatic readiness is all and consciousness is idle. But Nature may instead, fortuitously, have hit upon a shortcut requiring far less advance "planning," far fewer teleological schemes proleptically realized in advance of any given goal-directed behavior: Instead of many different schemes, each workable in one possible set of forthcoming circumstances, Nature may have produced one scheme not obviously workable under any conditions and conscious readiness with enough flexibility to bend it to fit many different sets of circumstances, with enough power to implement the bent scheme effectively.

One could argue in favor of the latter hypothesis on the grounds of its simplicity, but Nature may be more profligate in its ways—especially with respect to evolutionary processes—than such an argument would presume. Besides, even if Nature were parsimonious in the extreme, the argument would lack a metric for assessing relative complexity. Multiplicity of proleptically realized schemes is not the concrete indication of complexity one might think: An artificial automaton programmed to function in accord with a recursive procedure may be said to realize proleptically all the vast numbers of outcomes generable by way of that functioning; but the complexity required of this automaton might pale by comparison with that required of one programmed to follow a far less efficient, not so productive procedure. And in the case of a putatively natural automaton, an animal organism, it is not even so obvious that proleptically realized schemes reside within it (as opposed to, say, within the whole epigenetic system—the changing organism and its changing environment), so there is still less reason to assess the automaton's complexity on the basis of the sheer number of those schemes. And, finally, for all one knows, the emergence of conscious readiness may demand underlying neurologic complexity of a magnitude well beyond that of a teleological Automaton whose plasticity would enable it to pass Nature's survival test—a variant, as it were, of Turing's test.

But even if the argument from complexity is unsustainable, there are still some things (of an empirically speculative sort) to be said in evolutionary favor of flexibility rather than teleological plasticity: The best laid plans gang aft agley, and rigid adherents to those plans are apt to go astray with them; so, we may entertain some real (not merely philosophic) doubts that a teleological Automaton could pass muster. Then, too, were Nature to have produced genuine flexibility, Nature could really be as unconstrained as it appears in its further random anatomical experiments: If flexibility were available to make suitable adjustments during ontogeny, Nature's sportive variations—say, in length of limbs—would not so much demand, for survival's sake, any parallel, coordinated phylogenetic variations of teleological schemes for vital goals. Of course this consideration might be thought to argue just as well for the nativistic suggestion that Nature is simply more extravagant in its phylogenetic production of organism-resident schemes than some instinct theorists might suppose, that Nature has, for example, devised automatistic intelligence to generate them as needed. But without denying the possibility of such extravagance, I still wonder if it could wisely be managed mechanically. I have my doubts, that is to say, about the prospects for automatistic intelligence; and, in the absence of anything more than confident claims about it, I am not prepared to chant along with true believers in its existence.

They and others of the deterministic persuasion are dismissive in kind of any suggestion that there might be some exception, some reasonable opposition to the rule of well-established mechanical procedures or processes. Does it make sense even to suggest that we conscious creatures of Nature have any real control over our own destinies? In a normal frame of mind it might seem more absurd to suggest we do not, but a consciousness affected with scientistic religiosity cannot think straight. The distemper is not science but religion.

10.3 Interrogating Bergson's Argument for Vitalism

The position I espouse may fairly be adjudged a type of vitalism—not the belief in a general life force, but the suggestion of a partial self-determination by some living things of their purposive behaviors. Both types are accused of mysticism, as if to say that no good reasoning could possibly tie vitalism to reality. Instead of accepting this accusation on faith, let us examine Henri Bergson's argument for the second sort of vitalism. Bergson's version of this vitalism is not quite what I would endorse, but his premises are agreeably reminiscent of some (previously discussed) views of Dewey and Kuo:

> The cerebral mechanism is arranged just so as to drive back into the unconscious almost the whole of [our] past, and to admit beyond the threshold only that which can cast light on the present situation or further the action now being prepared—in short, only that which can give *useful* work.

> Doubtless we think with only a small part of our past, but it is with our entire past, including the original bent of our soul, that we desire, will and act.

> From this survival of the past it follows that consciousness cannot go through the same state twice. The circumstances may be the same, but they will act no longer on the same person, since they will find him at a new moment of his history. Our personality, which is being built up each instant with its accumulated experience, changes without ceasing.

> Thus our personality shoots, grows and ripens without ceasing. Each of its moments is something new added to what was before. We may go fur-

ther: it is not only something new, but something unforeseeable. Doubtless, my present state is explained by what was acting on me a moment ago. In analyzing it I should find no other elements. But even a superhuman intelligence would not have been able to foresee the simple indivisible form which gives to these purely abstract elements their concrete organization. For to foresee consists of projecting into the future what has been perceived in the past, or of imagining for a later time a new grouping, in a new order, of elements already perceived. But that which has never been perceived, and which is at the same time simple, is necessarily unforeseeable. Now such is the case with each of our states, regarded as a moment in a history that is gradually unfolding: it is simple, and it cannot have been already perceived, since it concentrates in its indivisibility all that has been perceived and what the present is adding to it besides. It is an original moment of a no less original history.[8]

Understood as suggesting that our consciousness together with existing circumstances determines our future state in accord with our present state (i.e., our personality as formed by our entire past, including our "original bent"), Bergson would seem to be arguing that the past's invisibility and indivisibility in present consciousness precludes any prediction of our future course. There is some plausibility to this argument, if viewed as directed against the prospects for predicting our future state on the basis of simple induction from our past as given in the immediate data of our present consciousness. But why limit ourselves, other observers, or even an omniscient being (God) to this particular procedure? We might think instead to ignore consciousness altogether and base our prediction on the past, on our developmental history, as it lends itself to objective scrutiny. Then, whether or not the pattern of development discerned were amenable to further analysis by (what Kuo calls) "atomistic methods," this total pattern might arguably (see section 4.5) afford a basis for sound prediction.

Or are we, by ignoring consciousness, failing to include both a principal determinant of our future state and a principal constituent of that state? We may stop worrying about the latter possibility simply by resting content with predicting only those aspects of our future state that are also amenable to objective observation, but the former possibility still provokes a question: If consciousness is a principal determinant of objective as well as subjective aspects of our future, could anyone, even God, hope to predict objective aspects of that future without taking account of that determinant?

A compelling affirmative reply, which might seriously undermine Bergson's position, requires some philosophical preliminaries: One plausible contemporary view of the mental would have it that states of consciousness "supervene" on the physical—which is to say, that they depend for their existence upon and are uniquely determined by specifically physical features of reality.[9*] Insofar as this view is conceived to be essentially materialistic (rather than dualistic), the dependence averred must be such that in the absence of all the sets of physical factors, each set of which uniquely determines a state of consciousness, there would be no (state of) consciousness. The hypothesis of such absolute dependence on these sets—known (collectively) as a "supervenience base" [10*]—effectively precludes Cartesian dualism, which requires the incompatible hypothesis that consciousness might survive the destruction of the material world. But despite its anti-Cartesian import, this view of consciousness as supervenient upon physical factors is quite compatible with the suggestion that consciousness is not just a logical construction built entirely of those (individual) factors, but a genuinely emergent aspect of the organism-in-environment to which those factors pertain. (This compatibility claim is premised on the general truth that emergent aspects of anything could not endure that thing's loss—much as the smile on the face of a non-Chesire cat could not survive that cat's disappearance.)

Given the suggestion that consciousness is an emergent thus supervenient upon certain physical factors, it might be argued that even if consciousness is a principal determinant of a person's objective future, accurate predictions of that future might be possible in ignorance of the (invisibly indivisible) form of the person's consciousness: For if consciousness varies uniquely with those (presumably objective) factors, and a person's objective future varies uniquely with consciousness plus objective circumstance, then that future varies uniquely with objective factors neat—thereby allowing in principle for the prediction of a person's future on the basis of objective conditions alone.

This anti-Bergsonian possibility of prediction may be made plain by resorting to a fanciful philosophical dodge: Suppose God decides to conduct some fantastic "thought experiments," ones involving the concrete creation of possible worlds. God conceives an exact duplicate of this world, then waits a moment to see whether creatures identical to us in all observable respects, including historical ones, will behave exactly as we do. As it happens, these other creatures always do behave as we do. So, whenever God wishes to predict what we will do in a given situation, God conceives a duplicate world in which we face that situation, then waits to see what our doubles do. [11*]

This imaginary Godhead experimentation further discredits the Bergsonian lemma that our future behaviors are unforeseeable even by God. His larger argument is that, given their unpredictability, our behaviors prove to be inherently novel creations. Yet, as the Godhead experiments also show, the sense in which Bergson deems the behaviors novel (or original) is not the sense implicit in the notion of anti-instinctive flexibility; for his sense is compatible with the idea that our behaviors are uniquely determined by our (world-specific) unique history, and this idea entails that our behaviors were destined to be, that their schemes were prefigured in initial prolepses.

Indeed, Bergson's remarks seem not just compatible with this idea but in its service. His vitalism with respect to behavior seems a deterministic variety quite favorable to instinctivity: Purposive behaviors are judged self-determined (by agents) because, given the uniqueness of an action's determining conditions, of each agent's original bent and its historical unfolding, there is nothing else these behaviors could be said to be determined by; but despite this uniqueness, which renders foresight problematic, the behaviors are uniquely determined by their (causative) preconditions. If this interpretation of Bergson is correct, then the failure of his unpredictability lemma is not a serious blow to his basic position: That lemma arguably remains true in this world, from any vantage point within it; and the predictability from a divinely transcendental perspective just lends substance to the deterministic element in his vitalism, by considering (in pragmatic fashion) what conceivable difference that determinism makes. So although Bergson's argument may fail to establish it, his position is not wholly untenable.

10.4 A Denial of Supervenience: Neutral Monism and Psychological Indeterminacy

Although my talk of possible worlds might raise some philosophers' hackles, it does not result in any new ontic commitments. To lower the hackles, I might align myself with the position that this talk is just a vivid way of speaking as it were, speaking subjunctively, about this world—the real world. But one legitimate use of the subjunctive mood is to avoid committing oneself to the existence of what one talks about, so I may as well, to equally good philosophical effect, speak subjunctively of the possible worlds themselves—by saying, for example, that were there other worlds (of the specified sorts) conceived by God, then such and such would/might occur.

The real philosophic danger here is not of ontic overcommitment but of overconfidence about happenstance in the other worlds conceived. The outcomes of the imagined experimentation are not foregone conclusions: God could discover, contrary to anyone's deterministic expectations, that duplicates of us in identically fashioned worlds do not invariably behave as we do and, so, that God is quite unable to predict the behavior of conscious creatures.

But what would account for such a turn of events? Some general indeterminacy with respect to the behavior of subatomic particles might suffice, but I am after options specifically favorable to vitalism with respect to purposive behaviors, so I will set this possibility aside as too general. That leaves at least two lively options especially pertinent to consciousness as a determinant of unpredictable actions: First, consciousness, although a determinant of behaviors, is not uniquely determined by objective conditions; second, consciousness (combined with circumstance) determines behaviors nonuniquely.

Since the two options seem logically independent, let us consider one apart from the other. Despite its denial of supervenience (i.e., its repudiation of unique determination by the physical), the first option need not be understood to disassociate consciousness from physical reality; indeed, the option is consistent with—and will hereafter be understood in terms of—the antidualistic suggestion that consciousness is an aspect of the same reality of which the physical is another. Before considering the full option, let us examine Bertrand Russell's version of just this suggestion. Russell's account stops short of the first option, however, for the account does not credit consciousness with being a determinant of behavior.

Russell, in *The Analysis of Matter*, claims that mental events "are part of the material of the physical world" but does not claim "that all reality is mental." [12] He regards physics as "supreme among the sciences," [13] owing to the exactness of its laws, and sees its aim as discovering "the causal skeleton of the world." [14] This skeleton (which, he says, "physics seems to prove that there is" [15]) is a metaphor for a species of determinism according to which the world's most basic underlying causal structure is a closed system of deterministic (i.e., uniquely determining) physical laws. Since the qualitative properties of mental events—doubtless including states of consciousness—are not among the exclusively structural properties mentioned in physical laws, Russell concludes that even when the structural properties of certain events, effects, are uniquely determined by physical laws, "these effects could be qualitatively of different sorts." He continues: "If that were so, physical determinism would not entail psychological determinism. . . . This is an unavoidable consequence of the abstractness of physics. If physics is concerned only with structure, it cannot, *per se,* warrant inferences to any but the structural properties of events." [16]

These remarks do contest, as does the first option, the supervenience-theoretic notion that states of consciousness are uniquely determined by physical factors. But lest we suppose much freedom would result from the psychological indeterminism alluded to, Russell invokes the fateful skeleton, physical determinism:

> Even if we reject the view that the quality of events in our heads can be inferred from their structure, the view that physical determinism applies to human bodies brings us very near to what is most disliked in materialism. Physics may be unable to tell us what we shall hear or see or "think," but it can, on the view advocated in these pages, tell us what we shall say or write, where we shall go, whether we shall commit murder or theft, and so on. For all these are bodily movements, and thus come within the scope of physical laws.[17*]

Now to secure the full first option I proposed, it is necessary to go beyond the psychological indeterminacy whose possibility Russell concedes, to give consciousness a subsequent causal role in the production of behaviors. Yet if the existence of a causal skeleton is as firmly established by physics as Russell seems to think, how could consciousness possibly play that role and thereby render the behaviors unpredictable?

The obvious answer is that consciousness might help determine (perhaps uniquely) the nonstructural properties (or qualities) of behaviors whose purely physical properties were utterly predictable. But, then, what kind of qualities could effectively stymie God's awesome power to predict the behaviors? Consider the qualities of sensibleness sometimes ascribed to behaviors. If, as Ryle suggests, "there need be no physical or physiological differences between the descriptions of one man as gabbling and another talking sense, though the rhetorical and logical differences are enormous,"[18] then God's otherwise accurate predictions of objective properties of behaviors might fail to predict objectively discernible differences in the sensibleness of those behaviors. But is Ryle's suggestion tenable? Here the supervenience theorists' slogan "No difference without a physical difference"[19] comes into its own, for surely there could be no difference in the reasonableness of two behaviors without some physical differences between them. Thus, saying "2 + 2 = 4" would be physically different from saying "2 + 2 = 5." If Ryle's point were that the sensibleness of physically identical behaviors might differ solely in consequence of different contexts, then even without contesting the point it could be observed that different contexts would themselves be physically different. And if the point is, rather, that differences in the intelligence of performances might be based solely upon dispositional differences, upon differences in what two otherwise identical performers "would do if," then his position runs afoul of the causal skeleton (which precludes their doing anything physically different).

Only an archdualist would countenance the extravagant idea that physically identical behaviors of physically identical organisms in physically identical circumstances might be differently intelligent—owing to purely mental differences behind the scenes; but instead of rendering any objective properties of behavior unpredictable to God, this idea merely denies that intelligence (or reasonableness) is such a property.

Russell is surely right to see the causal skeleton as imposing tremendous constraints upon the range of potential psychological indeterminacy. Perhaps only purely subjective, physically nonefficacious qualities might vary independently of physical properties, for different objective qualities would seem bound to elicit divergent physical reactions (e.g., "Oh look, this behavior is slippish, whereas the other is widgy"), and, given the closed character of the skeletal system, this could only happen insofar as those qualities were associated with different, differently efficacious physical properties. But if the only physically underdetermined qualities were subjective, then God's discovery of unpredictability with respect to them would be rendered inconceivable—unless God could (which I doubt) be conceived to be immune to the notorious philosophical difficulty of fathoming others' subjective experiences. So the prospects for unpredictability remain bleak, and the skeleton seems to grin at our predicament.

10.5 Russell's Causal Skeleton:
Destiny and Hierarchical Determination

If unpredictability of objective features of behavior is a requirement of true freedom, then Russell's dual-aspect doctrine of one reality, his "neutral monism,"
cannot be used to reconcile physical determinism with true freedom. Freedom
may require our challenging physical determinism, but before we do, let us consider what valuable lessons about the character of physical fate might be learned
from Russell's account.

A fuller account of the world as studied by science has it that the world is arranged into hierarchical levels of organized complexity, where each higher level
would seem to include some "emergent phenomena" not contained among any
less complex (lower) levels. Corresponding to each level is at least one science or
scientific subspeciality concerned to study that level—if only because phenomena
are not considered integral to a distinct level unless they have seemed to demand
special scientific explanation. But even when or if special laws or theories do
explain the goings-on at higher levels, it might still be possible to explain the
explanation—to "reduce" the higher-level laws or theories to lower-level ones.

Given the widespread belief in physical determinism, it seems incongruous
that the most basic level(s) is (are) commonly held, on good empirico-theoretical
grounds, to involve rampant indeterminacy, to be governed, for example, by
"probabilistic propensities" rather than exact causal laws. But these discordant
beliefs are harmonized by suggesting that determinism is instituted at a higher
level of the hierarchy—as if perhaps some ill-coordinated individual members
were to constitute a precision team. Once it takes hold of things, physical determinism is presumed never really to relinquish control of them: All higher levels
abide by the deterministic laws of lower levels. Exceptions to the rigid rule of these
laws are admitted, but only insofar as lower-level (micro)indeterminacy manifests
itself at macroscopic levels—by magnifying its effects upward (as may be presumed to happen when physicists use sophisticated devices to detect phenomena
indicative of microindeterminacy).

Russell's slant on how physical determinism comes about is to suggest that
its laws are really just "statistical averages [sufficient] to determine macroscopic
phenomena"[20]—phenomena above the level where indeterminacy "rules." This
suggestion mixes levels and should probably be revised to say that deterministic
laws at $level_n$ are reducible, somehow, to statistical averages at $level_{n-1}$. Russell's
words might lead us to mistake the situation as a matter of $level_{n-1}$ events causing
$level_n$ events, but I take his point to be, rather, that the apparently deterministic macrolevel laws are really just statistically (very) reliable average tendencies
at the microlevel, that what is going on at the macrolevel is in fact identical to
something taking place at the microlevel.

On the assumption that there is only one hierarchically arranged world, it is
tempting to think that everything at the higher levels is identifiable with something

at the very lowest level—with some grouping(s) of "ultimate constituents" of the microphysical world. But this simplistic building-block conception of reality does not follow from the assumption, as a Russell-inspired understanding of the hierarchy should make plain. What needs emphasizing is that emergent phenomena are those that simply do not figure in scientific frameworks (theories, etc.) associated with less complex levels, that emergents are science-relative (or theory-relative) novelties. Accordingly, there is no reason to think that emergents are identifiable with any purely physical groupings characteristic of the levels at which those emergents appear. Although the conceptual connection between levels and complexity might seem to support a building-block conception of reality, it should be noted that complexity is here relative to a particular science, physics, and that, notwithstanding any "bridge laws" between types of complexity and types of emergents, the emergents themselves lie outside the (structural property) domain of that science.

There is surely much truth to the idea that physics is the most comprehensive science, but even this truth cannot be used to deduce the truth of the building-block conception. The claim of comprehensiveness is not just that physics studies more of the world as spread out in time and space than does (say) biology, for by the same token biology studies more of the world as living than does physics. The claim is best understood (apropos of the same two sciences) in aspectival terms: Physics studies all living things in their physical aspects; biology studies all physical things in their living aspects; but physics is more comprehensive than biology, because even though all living things are physical, not all physical things are (so far as we can tell) living. Analogous remarks may be made in comparison of any other science with physics, and thereon rests its claim to be the most comprehensive of all sciences. But such comprehensiveness is logically compatible with a denial of the (physical) building-block view: Everything in the world may have its physical aspects, but not every aspect of everything is in consequence physical.

This compatibility gives us philosophic license to hypothesize that physics is comprehensive but not exhaustive, that in fact not all aspects of reality are physical. Since this conjecture is prompted by Russell's neutral monism, it ought to be in keeping with his understanding of physical determinism. But what about his proposed reduction of that determinism to microphysical statistical averages? The conjecture should make one wary of some reductionistic moves, but it is no general cause to worry about cross-level identifications of strictly physical aspects of the world. One might use the conjecture as a basis for concern about saying, reductionistically, that $level_x$ physical phenomena are really just $level_{x-1}$ physical phenomena; but in Russell's case some such downward bias (in favor of lower over higher levels) would still seem warranted by the logical need to choose between determinism and indeterminacy and the implausibility of saying, for example, that indeterminacy is really just determinism writ small.

The downward bias, which manifests itself in several ways, may be an outgrowth of a building-block conception, but this conception, suitably revised (i.e., restricted and relativized) to say that some things at each higher level are identical to some things at each lower level, has residual philosophic merit—how else explain higher levels as increasingly complex? And there is some perhaps not wholly question-begging appeal to the idea that this bias is an understandable preference for more (over less) fundamental explanations. Nevertheless, the merits of the bias should be assessed case by case.

Consider the presumption that lower-level events may cause higher-level events but not vice versa. If there is a truth in this downward bias in favor of upward causation, it seems badly expressed. One radically different way to put it would go so far as to deny all causal relations between levels, claiming instead that only intralevel causal relations obtain. The bias itself might then be expressed as a suggestion that the level at which these relations actually obtain is invariably the lowest among all at which they seem to obtain. Behind all this is the further thought that if any higher-level events are reductively identifiable as, are "really only," complexes of lower-level events, then whatever happens at the higher level is not so much caused by lower-level events as constituted of them. The changes on high are, as it were, already wrought below—though any hint here of temporal priority is literally false: Events below uniquely determine the changes above, but this is an atemporal (hence noncausal) determination.

This expression of downward bias helps to explicate Russell's notion of a causal skeleton: 1) Physical causation is initially predicated of $level_n$, which is characterized in exclusively physical-structural terms. 2) The domain D of $level_n$ events is closed with respect to causal relations: For all x and y, if x is a member of D, and x is a cause of y or y is a cause of x, then y is a member of D. 3) All physical causation at $level_{n+1}$ or above is reductively identifiable with (or analyzable in terms of) causal relations among $level_n$ events.

This account of physical determinism still rings a bit hollow, since microindeterminacy reigns at $level_{n-1}$ (and below?), and $level_n$ phenomena are all reductively identifiable with $level_{n-1}$ phenomena. But let us not plumb the depths lower than $level_n$. Russellian physical determinism does not preclude nonphysical causal relations from emerging at any higher level but does render them physically ineffectual everywhere. It might be claimed that physical causes are rendered just as nonefficacious with respect to nonphysical emergents, but this claim should be qualified since events involving emergents may be constrained even though not caused by physical events. There are varieties, not all of them causal, of event determination. The type just alluded to will be termed *cohibitive determination* (or *cohibition*).

Russell thinks it "probably true" that "we could infer the qualities of the events in our heads from their physical properties."[21] This thought suggests that the mental aspects of events are uniquely cohibited by their physical properties—

a suggestion that implies but is not equivalent to the claim that mental properties supervene on physical ones. The claim of supervenience (as this notion was characterized above)[22*] also allows for the physical events' either causing or constituting mental ones, while the claim of unique cohibition specifically precludes the former alternative. For the sake of that curious philosophical clarity that sometimes springs from terminological complexity, let us mark the two alternatives by referring to the former, temporal variety as *causal determination* (or *causation*) and the latter, atemporal (set-theoretic) type as *constitutive determination* (or *constitution*).

The use of these terms is instanced by an earlier suggestion now paraphrasable as the claim that there is no interlevel causation, that what passes as upward causation is more properly described as the constitution of higher- by lower-level events among which "real" causal relations obtain. In sharp contrast to this claim is the idea that not only is interlevel causation a possibility, but downward physical causation is a reality. The truth of this idea would seem inconsistent not only with that claim but also with Russellian physical determinism.

10.6 The Closed System and Its Enemy: Popper on Downward Causation and Deep Indeterminacy

Karl Popper, in defense of the view that consciousness is an emergent determinate of bodily movements, argues for (what D. T. Campbell had called) "downward causation."[23] Popper presents some supportive physical cases: "A diffraction grating or a crystal . . . is a spatially very extended complex (and periodic) structure of billions of molecules; but it interacts as a whole extended periodic structure with the photons or the particles of a beam of photons or of particles. Thus we have here an important example of "downward causation.". . . That is to say, the whole, the macro structure, may *qua* whole, act upon a photon or any elementary particle or an atom."[24] Also: "Stars are undesigned, but one may look at them as undesigned 'machines' for putting the atoms and elementary particles in their central region under tremendous gravitational pressure, with the (undesigned) result that some atomic nuclei fuse and form the nuclei of heavier elements; an excellent example of downward causation."[25]

Both cases are compelling, but it could be argued that they do not really involve interlevel causation since purely physical terms suffice to describe all causally pertinent aspects of each case. Insofar as "higher level" presupposes emergent properties at a level, the argument has a point; but if reducibly emergent properties, that is, ones constitutively determined at a given level, will serve to distinguish a higher level, then the point of the argument seems lost (since then the cases' describability in exclusively physical terms will not obviously belie the claim of interlevel causation).

Both cases may be described in identical general terms as involving special effects being wrought in low-level items by dint of a spatio-temporally nearby complexity. Notwithstanding that this complexity is constituted of items at the same low level as the ones affected, Popper contends that the whole is here differently efficacious than the mere sum of its parts would indicate. Whether he is right, only science could, perhaps, tell us; but let us just assume that he is. This suggests that the domain of level$_n$ events is not closed with respect to causal relations, though it does not imply that there is anything wrong with what the physical theories tell us about some causal relations at level$_n$—namely, those obtaining among level$_n$ items not in the causal vicinity of any complexity. (Where a causal vicinity is the collective spatio-temporal extent of causal relations to every item in the subdomain.) Indeed, one might imagine another, less complexity-ridden world for which those theories held entirely true. Such a world would, despite its nomological-constitutive similarities to our own, exhibit the level$_n$ causal closure that ours, by Popper's reckoning, lacks.

Russell's causal skeleton begins to loom less formidably. The causal closure ascribed to all level$_n$ events may in truth, in this world, be only a de facto characteristic of some isolated subdomains of level$_n$. Should conditions change, should other complex subdomains enter—indeed, extend—a subdomain's causal vicinity, then that once closed subdomain will be open. This is not far from what is meant when speaking of, say, a closed mechanical system: We do not suggest that such a system is immune to outside causal influences, only that in its present whereabouts, the system is (relatively) unaffected in its mechanical functioning by external forces.

Physics' comprehensiveness combined with the fact that each of its level$_n$ laws uniquely determines certain events, given others, does not establish the existence of a causal skeleton. It does not—tempting though it may be to argue that the situation described is akin to that of a closed mechanical system whose autonomy could not possibly be diminished by changes in conditions external to the system, as nothing exists outside the system. To see just what goes awry, let us make an amicable attempt to formulate the argument more precisely. The task is facilitated by making some simplifying assumptions—namely, that (physical) events take place in Euclidean four-dimensional space and that events are spatio-temporally contiguous with their causes and effects.

The argument requires some definitional preliminaries: The level$_n$ system consists of all those level$_n$ events that are causally connected to other level$_n$ events. This system, R_n (for "Reality at level$_n$"), is a subset of the set (or space) of all spatio-temporal events. The distance between any two events, e_1 and e_2, in R_n is defined in terms of their respective spatial and temporal coordinates, (x_1, y_1, z_1, t_1) and (x_2, y_2, z_2, t_2), thus:

$$d(e_1, e_2) = [(x_1-x_2)^4+(y_1-y_2)^4+(z_1-z_2)^4+(t_1-t_2)^4]^{1/4}.$$

Given this distance function, R_n is a Euclidean, hence metric, space.[26*] Now the (causal) neighborhood, N_n, of each event, e_n, in R_n is the set of all events (or points) that are less distant from e_n than is the cause of the cause of e_n. Where e is a point of R_n, e is an interior point of R_n if a causal neighborhood of e consists exclusively of points of R_n. If the causal neighborhood of e contains both at least one point of R_n and at least one point of its complement, $C(R_n)$, e is a (causal) boundary point of R_n. Finally, R_n (or any other metric space) is open if and only if all its points are interior, and closed if and only if it contains all its boundary points.

Given these preliminaries, the argument for the causal skeleton can be interpreted to conclude that R_n is closed. From the premise that R_n is the whole world (or whole space), so that $C(R_n)$ is the empty set, it follows that the set of all the boundary points of R_n is empty. And since, by convention, the empty set is a subset of every set, the boundary of R_n is contained in R_n. In other words, R_n is closed.

This argument might be questioned for failing to provide a noncontextual definition of the distance involved in specifying (causal) neighborhoods and, indeed, for tacitly assuming that all events spatio-temporally contiguous with a given event are equidistant from it—that is, are the same distance from that event as is its own cause. But let us give the argument the benefit of these doubts.

Even so, we could go on to quibble with the argument by observing that if the world were nothing but causally connected $level_n$ events, that world, R_n, would be not only closed but also open—since all of the points of the whole space R_n would be interior points. But this sort of openness is not helpful to the case against the causal skeleton, since it does nothing to create an internal route of escape from that skeleton.

The real weakness of the argument for the skeleton is the quite unwarranted assumption that $C(R_n)$ is empty. The system R_n of $level_n$ events need not be the whole world, even granting that those events occupy all spatio-temporal regions occupied by anything of causal consequence. It is still possible for there to be internally "external" events, some emergent phenomena in the proximity of (lower-) $level_n$ events and causally efficacious with respect to them. The existence of such emergents would negate a crucial premise of the argument for the skeleton: The complement of the $level_n$ system would not be empty; $C(N_n)$ would contain higher-level spatio-temporal events.

Since the physical emergents proposed by Popper do occupy the same spatio-temporal locations as the equally physical $level_n$ events, those higher-level events would seem uniquely constituted by the set of lower-level ones. How, then, could these emergents possibly be a complementary set of $nonlevel_n$ events? One quick answer, which may not go much beyond the question, is that the emergents have complexity and other properties of a whole that are not intelligibly ascribed to its parts. A more considered response, which may go well beyond the question but not fully address it, is that the emergent events have, in consequence of such properties, differently efficacious tendencies. This response may seem to beg other questions, larger philosophical concerns about emergents as metaphysical extras, but it does accord with what the scientifically hierarchical conception of the whole world has to say about higher levels and their apparent emergents—namely, that they must seem to demand special scientific explanations. Popper's remarks imply that, in the cases he mentions, this demand will not go unmet and that the theories used in the explanations will not be reducible to lower-level theories.

Popper's position seems most opposed to a particular construction put on determinism—namely, that $level_n$ is a closed causal system of physical events that uniquely constitute all physical aspects of higher-level events. But the reconstruction of Russell's causal skeleton is not laid waste yet, even if we accept Popper's above claims as gospel. Downward causation may yet be a unique determination, and then we would have no reason to suppose that the physical world is on the whole—at $level_n$ and above—physically indeterministic. Since we are promised special laws that would license us to assert that a $level_{n+x}$ event caused a $level_n$ event, we might infer that Popper's claims do deny the causal skeleton hypothesis, but only if there is further reason to suppose $x > 0$. The only independent reason already given to suppose this is the complexity of the cause, but at $level_{n+x}$ it is the complexity qua whole that is presumed causally efficacious. Since complexity qua whole is just another structural property of the world, and since a deterministic law purportedly relates this complexity's occurrence to other $level_n$ events, the occurrence fully qualifies as a $level_n$ event—given that $level_n$ is, by definition, the first level where physical determinism takes hold.

To do more damage to the causal skeleton, Popper must dispel the idea that his emergents belong to $level_n$—their holistic complexity may be a requirement for, but is not proof of principal residence on higher ground. Apropos of another emergent, heat, Popper observes that its explanation is incompletely reducible to explanations in terms of motion, atoms, molecules; "for new ideas have to be used—the ideas of *molecular disorder* and of *averaging;* and these are, indeed, ideas on a new holistic level." He then proposes another case of downward causation, a case involving heat and explained via such ideas: "The *average* velocity of a *group* of atoms influences the *average* velocity of the neighboring *groups* of atoms [and] thereby influences . . . the velocities of many individual atoms in the group." The case is presented as pertaining to the heat of a crystalline solid, the cogwheel of a clockwork—emblematic, for him, of mechanistic determinism; and the downward influence is said to be "due . . . to the random character of the heat motion, and therefore, I suspect, to the cloudlike character [emblematic of *indeterminacy*] of the crystal." [27]

One sure reason to deny $level_n$ membership to a phenomenon would be its introduction of indeterminacy into the course of physical events. And were such a phenomenon an emergent explicable only by holistic ideas unavailable at explanatory levels lower than $level_n$, that phenomenon's place on high would be assured. But Popper's proffered explanation does not satisfy the former condition: His implicitly probabilistic interpretation does not entail physical indeterminacy. Popper's shift of terms, from "downward causation" to "downward influence" (and then to "dominant influence upon a lower level"[28]) may prepare us, rhetorically, for his apparent conclusion-cum-suspicion that the effect on the velocity of an individual atom is not uniquely determined by the higher-level phenomenon heat. But the fact that the effect is held to be only probable should not, by itself, persuade us of the truth of that conclusion, for sometimes probabilistic explanations merely betray imperfect, not unattainable (for example, by God) knowledge.

Popper suggests that "were the universe *per impossible* a perfect determinist clockwork, there would be no heat production and no layers and therefore no such dominating influence would occur." And though itself an exaggeration, this suggestion leads him to note, correctly, "that the emergence of hierarchical levels or layers, and of the interaction between them, depends upon a fundamental indeterminism of the physical universe."[29] Yet this point is not as significant as it may seem: All it can mean is that, in the absence of indeterminacy (or, lack of causal closure at some purely physical level, call it $level_n$), all would-be interlevel physical causation may be reinterpreted as $intralevel_n$.

The great strength of the causal skeleton is not its iron necessity at a predesignated level of physical complexity but its conceptually conferred ownership of anything physical, however complex, that forms part of a closed system governed by laws that uniquely determine its future (on the basis of past) states. Popper overstates when he suggests that physical determinism precludes any emergent phenomenon, be it felt heat or averaged atomic velocities, since neither need vanish in order to be incorporated, physically accommodated, within the skeleton.

Popper may also exaggerate by hinting that all indeterminacy is a consequence of low-level indeterminacy (i.e., the imperfectly predictable cloudlike character of what we mistake for utterly reliable components of a clockwork universe). The assumption that all indeterminacy dwells at microlevels of physical complexity is, so far as I can tell, a quite unwarranted downward bias; and once it is revealed as such, there seems no reason not to allow that some indeterminacy might emerge at higher levels and remain there. A closed system may bring all of causal consequence to it down to one level, but an open system need not so debase its members.

Of course it might be argued that if ultimate constituents *uniquely constitute* all levels of complexity, then higher-level indeterminacy must be wrought below, as an atemporal function of those constituents. But indeterminacy is hardly the prior constraint that determinism is: An emergent indeterminacy at a particular level of complexity might not be *uniquely cohibited*, even if the complexity were thus constituted; and, even if the emergent were thus cohibited, it would not in consequence follow the lower level's lead. Higher-order indeterminacy is not, by nature (or this characterization), a mere pawn of any lower-level regularity much less a victim of still lower-level inconstancy.

Higher-order indeterminacy might best be viewed as cohibitively determined by lower-level events and as an emergent phenomenon in its own right. Were we to reduce all higher-order to lower-level indeterminacy, then the freedom sought on high (in consequence of emergent consciousness) would be at the beck and call of randomness.

10.7 Consciousness Naturalized: Supervenience without Physical Determinism

It is time to recall the two options available to account specifically for the inherent unpredictability of consciousness-wrought behaviors: Indeterminacy might enter either into the production of the consciousness or into its production of the behaviors. An exploration of the first option has led, via an extended discursus on the nature of physical determinism, to the present point in the argument. If physical determinism is maintained, the first option is still available; but it then affords little in the way of meaningful freedom. Indeed, it affords nothing in the way of objective unpredictability, since all physical aspects of the behaviors would be a foregone consequence of other physical aspects of the world. One might choose not to uphold physical determinism, to reject the causal skeleton hypothesis. This choice could in good philosophical conscience be made, and for the sake of meaningful freedom would seem philosophically de rigueur. The choice, simply put, is to deny that the physical world is a causally closed system.

Nonphysical aspects of the world, including occurrences of consciousness, might then be in a position to be causally efficacious with respect to the physical aspects of the world. The very idea of nonphysical aspects of the world that are not uniquely determined by its physical aspects does deny some strong philosophical intuitions about the world, intuitions currently expressed in the doctrine that all aspects of the world are supervenient upon—that is, uniquely determined by and absolutely dependent on—its physical aspects. But against the backdrop of Russell's (dual aspect) neutral monism, the unique determination hypothesis may seem less compelling—even to a physical determinist, much less to a physical indeterminist. And if it is antidualism one is after, the hypothesis proves unnecessary—one only needs to assert (a version of) the hypothesis of the absolute dependence of nonphysical upon physical aspects of the world. This remaining core of the supervenience doctrine still supports the view that physics is the comprehensive science but does not for all that commit one to a naive building-block conception of the world.

But taking the first option without the second is still a philosophically suspect choice. It would have us attribute the resultant indeterminacy of behaviors to a settled state of consciousness, a state that simply has not been uniquely determined (at least not fully) by physical factors. This would not give consciousness itself any leeway, any flexibility of its own: A given state of consciousness would, in its physical aspects and circumstances, uniquely determine physical behavior. As a way of arguing against instinct, against fate, this leaves much to be desired. Bergson, who takes this path, does not repudiate instinct so much as further internalize it, subjectivize it as our innermost nature, our original but physically undetermined (or underdetermined) bent—unknown even to ourselves, save for its effects. Let us not be so quick to deny supervenience—specifically, consciousness' unique determination by physical factors—that we join Bergson back at destiny's door.

Of course it could be said of a state of conscious readiness whose teleological scheme is finally proleptically realized that that state is settled and that it, in its physical aspects and circumstances, uniquely determines behavior; and saying this is much to the good, for doing so should lessen philosophic concern about the causal efficacy of consciousness. But the point to be made, then, is that this settled state is settled by consciousness itself, that the scheme of this state was not destined in the overall circumstances to be the scheme attempted. And if this point is insisted upon, then the present suggestion ought not to be confused with Bergson's affirmation of the first option minus the second; for by exercising its flexibility, conscious readiness may be said (in compliance with the second option) to determine a behavior nonuniquely—even though the behavior is, subsequent to the final prolepsis, uniquely determined by consciousness and then existing circumstance. Remember, though, that final prolepsis is not, by and large, a separate phase of action, temporally well marked by a moment of full inception; that typically we do not in a moment resolve upon a whole scheme of action and then, straightaway, attempt the whole exactly, behaving as if mechanically implementing a set of programmed instructions. When (consciously) performing an action, we usually remain ontologically, if not psychologically, irresolute throughout its course of behavioral events: Each small part of still unfolding, still changeable schemes achieves final prolepsis as its implementation begins, rarely much before. Conscious readiness may be, over time, the guiding force in the unique determination of the behavioral events, but higher-order indeterminacy, flexibility, reigns alongside over much the same period of time.

We do need, for true freedom's sake, to take the second option if we take the first; but perhaps we have no good reason to take the first at all. The second option absent the first has the singular advantage (over other ways to avail oneself of at least one option) of being fully consonant with the doctrine of supervenience; so let us press this advantage and assert accordingly that consciousness, though uniquely determined by and absolutely dependent upon physical factors, determines behaviors nonuniquely. This constellation of claims will surprise some supervenience theorists, especially ones who suppose their doctrine implies that everything at the humanly accessible macrolevel of reality is predictable on the basis of deterministic physical theory. But this supposition is simply wrong, unless it further presupposes—begging the question—physical determinism. (How such theorists then expect to account for upwardly magnified effects of micro-indeterminacy passes understanding, but that is their problem.)

The consistency of the second-option/supervenience position is my concern. My earlier trichotomy among types of event determination will help to clarify the issues. The position does not merely suggest that physical events uniquely determine the physical aspects of consciousness, for this would be a trivialization of the supervenience doctrine—even a dualist might concede this on the grounds that there are no such aspects. In requisite accord with the full doctrine of supervenience, the position could be that lower-level physical events *uniquely constitutively determine* the physical complexity upon which the existence of consciousness depends absolutely. This is extremely plausible yet does not seem to say enough; this does not capture the idea that consciousness itself, not just its underlying physical complexity, is uniquely determined by physical constituents. But if this seems a problem simply because consciousness is understood as an emergent phenomenon, then it must be remembered that some putative emergents (e.g., heat as averaged velocities of atoms) are thus constituted. It may be observed in turn, however, that some emergents (e.g., heat as experienced) are not so constituted—at least not wholly—and that consciousness, though arguably (often or always) lacking in phenomenal qualities of its own, is not only like those emergents but integral to them.

To accommodate the understanding of consciousness as a constitutively irreducible emergent (unlike heat qua averaged velocities), the second-option/supervenience position might be interpreted to say that the lower-level physical events also *uniquely causally determine* consciousness. But this would seem to say too much, to attempt to explain how the occurrence of consciousness emerges. It is not so obvious that a causal account is altogether fitting and proper since, for one thing, the cotemporaneous occurrence of consciousness and its underlying physical complexity suggests an even stronger union than that of effect and cause. Consider, by way of something more than an analogy, the case of felt heat: Heat as averaged velocities may cause felt heat and so, perhaps, the physical constituents of the former could be said to effect the latter emergent phenomenon; but once we contemplate the physical factors directly underlying the felt heat, the idea that they cause it begins to seem an illegitimate abstraction: They are (or, are inseparable from) the effect.

To preserve the aspectival intimacy of consciousness and its underlying physical complexity, we had better not say that the latter causes the former. Indeed, moreover, insofar as effects are logically distinct from their causes, any talk of causal relatedness would seem to allow for the logical (Cartesian) possibility that consciousness might exist in the absence of the physical complexity. And though the claim of absolute physical dependency does forestall this sorry dualism, it is better not to start a tear than to do so and at once attempt repair.

Claiming less than that a physical congeries causes consciousness might not strengthen their union but might help to keep it intact. Let us take the second-option/supervenience position to be that lower-level physical events *uniquely cohibitively determine* the occurrence of consciousness. This makes no pretense of explaining how the latter event, the occurrence of consciousness, comes to be and does not falsely suggest that the former events precede the latter one in time. But does this interpretation say anything more than that the lower-level events constitute consciousness? Well, it does rather expressly concede consciousness' status as an emergent—given that cohibition is characterized as a noncausal determination of events involving emergents; but since some emergents are themselves uniquely constituted by (seemingly) lower events, we have only a fool's assurance from this concession. Remember, though, that cohibition was also characterized as a matter of constraint as opposed to causation. This feature can go beyond mere constitution. In the context of neutral monism, cohibition might be understood as the curbed expression of some aspects of reality by others.

One silly version of this finds dual aspects all the way down—in the very form they have on high. Thus, it is suggested that atoms, monads, or other allegedly ultimate constituents of reality are conscious. The silliness stems from a fallacious application of the building-block conception to the case of consciousness; and it is scarcely an improvement to suggest that the ultimate constituents are potentially conscious—that is as bad as saying that the light pieces of a heavy object are potentially heavy. The core of truth is that there must be something about its parts that allows a whole to exhibit its emergent qualities—thus, the physical mass of light pieces allows for the heaviness of an object composed of them; and particular velocities of individual atoms allow for an average velocity of the totality of those atoms.

The precise parallel of this truth in the case of consciousness eludes us, but my confidence in that truth remains unshaken: Something there must be about the constituents of the physical complexity underlying consciousness that allows for its expression as constrained by this complexity. Now a reductivist might still insist, instead, that the (neurologically and otherwise organized) physical complexity exhibits higher-order physical indeterminacy and that that is all there is to consciousness (insist, i.e., that consciousness is uniquely constituted of purely physical ingredients). But I am inclined to suppose that there is more to consciousness than that, that it has emergent phenomenal aspects over and above any (also emergent) physical indeterminacy manifested by its physically complex basis. Something, I know not what, about the ingredients exhibiting the complexity and attendant indeterminacy allows for a particular quality of consciousness to be exhibited. Consciousness is an emergent aspect of a whole whose purely physical aspects uniquely determine it (consciousness) not constitutively but still cohibitively. (Constitutive determination is one but not the only type of cohibitive determination.)

Given this fuller explication of the second-option/supervenience position, a question might still be raised about its logical coherence. Does the possibility that consciousness may supervene upon the physical world but yet introduce indeterminacy into the production of behaviors not run counter to the supervenience theorists' slogan "No difference without a physical difference"? Not if that slogan is properly understood to assert consciousness' absolute dependence upon physical factors: A consciousness that exercises its flexibility may change only if the underlying physical factors that have (uniquely) determined that consciousness also change; but this does not imply that consciousness must abide by deterministic laws otherwise binding on physical factors—laws that mayhap be binding on all physical factors that do not belong to a supervenience base, do not uniquely determine states of consciousness.

What might lead one to suppose that some such implication holds is a faulty assimilation of unique cohibition to causation. To say that consciousness is (always) uniquely determined by a set of physical factors is to say not that they hold causal sway over states of consciousness, only that a functional relationship obtains between the physical factors in a set and a particular state of consciousness. If the relationship is further viewed as causal, it becomes tempting to view consciousness as bound by the deterministic laws presumed to have hegemony over physical factors determinative of consciousness.

Per coherent hypothesis, not even consciousness' underlying physical complexity is bound by those laws—its emergent indeterminacy transcends them, allowing for its hegemony over other physical factors in its causal neighborhood. Analogously (to some extent), our present consciousness is not an idle emergent suffering a fate determined by the organized complexity uniquely cohibiting that presence. Cohibition, unlike causation, can be and here is a two-way street. The ties that bind consciousness to its physical basis, its underlying organized complexity, are relations of mutually unique cohibitive determination. Consciousness and its physical basis are cohibitively equivalent dual aspects of the same organism in situ. The ontic locus of which the two are aspects is a proper subset not of the physical world but of the natural one, of Nature overall.

Chapter 11

Final Reckoning

11.1 The Dual-Aspect Nightmare

My proposed naturalization of consciousness qua determinant of behavior has obvious affinities to Spinoza's (now so-called) double-aspect theory, that "the decision of the mind, and determination of the body, are simultaneous in nature, or rather one and the same thing, which when considered under the attribute of thought and explained through the same we call a decision, and when considered under the attribute of extension, and deduced from the laws of motion and rest, we call a determination." [1] His one reality is also called Nature, though it's differently conceived to be more thoroughly imbued with mental characteristics: Each of its parts has a specifically mental (albeit not always fully mindlike) aspect as well as a physical aspect. There is another—related but more crucial—respect in which Spinoza's view differs from mine; he argues that free decision is illusive: "Whereof these decisions of the mind arise in the mind by the same necessity, as the ideas of things actually existing. Therefore those who believe, that they speak or keep silence or act in any way from the free decision of their mind, do but dream with their eyes open." [2]

This shared necessity is, on the physical side, deterministically nomological, pursuant to the "laws of motion and rest." But who is dreaming?

My fear is that someone will shove a frog-robot in front of me and ask whether or not it is a real, genuinely intelligent frog. Having made my guess, I will be told that either possible answer would have been correct: "It is a real, intelligent frog; it is also, we scientists have discovered, a robot—its brain, a computer." The episode is a nightmare. Pursuing the point with my tormentor, I ask: "Do you mean to say that it really is intelligent or only that it simulates intelligent functioning?" The scientist-cum-philosopher replies, "It all depends on how you look at it. I see it both ways: As a scientist, I'm all too aware of the underlying mechanisms that govern the frog's behavior. As a philosophically inclined member of *Homo sapiens,* I see the frog as a fellow (though sorely limited) member of the community of intelligent beings."

My reaction to this Spinozistic dream is that seeing-as will not make it so, that "you can't make a silk purse out of a sow's ear" just by seeing it as one. There is some fact of the matter that is not to be denied: Either the frog is or it is not gen-

uinely intelligent; and no amount of ingenious simulation, whether by AI theorists
or Nature or God, will warrant calling the frog intelligent, will justify seeing the
frog as intelligent, if it is not.

The fact of a frog's genuine intelligence would involve its conscious control
of its own seemingly savvy behavior. Such consciousness supervenes on certain
physically complex events in the natural order, though they take no ontological
precedence over consciousness and may equally well be said to supervene on it.
This emphasis on complexity may seem to put me more at odds with Spinoza
than with functionalistically inclined AI theorists, who care less about what a
system is made of than about whether it is complex enough to allow for certain
functioning; but reason for opposing them may be derived from a clearer sense
of the nature of my clash with Spinoza. My conjecture that some sort of physical
complexity is fundamental to a certain emergent physical indeterminacy, the ontic
dual of consciousness, is more a matter of philosophic caution than speculative
excess: We conscious beings are neurophysiologically, chemically, and just plain
physically complex organisms in a still more complex epigenetic system; of that
I am quite confident. But to reason from mental aspects of a whole system to
like aspects of its parts, which must somehow (I agree) afford that whole such
aspects, smacks of the fallacy of decomposition. I do not, however, attribute the
two emergent phenomena (higher-order indeterminacy and its dual, consciousness) to their basal physical complexity per se. That might well seem to begin
to favor some sort of psychological functionalism: Systems with the same complex organization of different parts—say, silicon-based rather than carbon-based
units—might then be expected to be functionally isomorphic and, so, alike in all
psychological respects. Instead, I make what I believe to be careful and judicious
appeal to my own ignorance, arguing that I do not know if not all chemical and
other constituents of the physical world may aspire to be part of a constellation
with dual (psychophysical) aspects; and I counsel like skepticism by others—
including AI theorists, who are thus well advised not to be too quick to admit all
able-bodied specimens as members of a "society of mind."

11.2 The Case against Psychological Functionalism

But my opposition to the functionalist runs deeper still, as does my opposition
to Spinoza. The disagreement turns, in both cases, on the possibility of higher-
order indeterminacy as a dual of conscious readiness. I will begin by disputing the
psychological functionalist. To meet this adversary on common ground, I will go
so far as to concede that any artificial or natural system functionally isomorphic
to a system that has conscious readiness may be judged to have it, too. In turn, I
demand the modest concession that the relevant details of the functioning of the
various systems may be characterized in terms of what I have called teleological

schemes; that is, abstract functional structures variously realized in the course of any truly teleological functioning, be it intentional or purely mechanical. One possible characterization might be an *output model* of a sort suggested by Turing's imitation game: teleological schemes abstracted from a system's teletyped responses to inquiries communicated to it via teletype. But further assurance is needed, then, that the schemes are true reflections of the system's own teleological functioning and not otherwise explicable, say, as interventions of a trickster who types in conversationally apt responses as artful translations of "what the computer 'really' meant to say" when it spewed out nonsense. Such assurance might be had by requiring, further, both that the schemes abstracted from the output be full realizations of *system resident* schemes—that is, ones already realized in the system's own final prolepsis—and that other plasticity or flexibility presupposed by teleological functioning be attributed to the system with respect to the schemes in its final prolepsis. But this requirement seems too stringent, since full realizations of teleological schemes—that is, overt realizations of schemes (in output) that exactly conform to ones in a final prolepsis—are not easy to come by, given the inestimable contingency of circumstance. What allows for the much greater frequency of occurrence of full realizations of computer (over organism) resident schemes is that the computer implements its schemes within its own "block universe," a reliable closed mechanical system consisting, say, of the computer and a printer. But not even an AI-theoretic psychological functionalist should want to adopt the output model if that means, for example, denying the functional equivalence of two machines that differ in their output solely in consequence of a malfunctioning printer to which they are both attached.

A more sensible characterization of functioning might largely ignore external manifestations of schemes, as realized in output, and emphasize instead their various system-internal realizations. This may—to note its congeniality to an AI-theoretic outlook—be called the *data-processing model* of functioning. System resident teleological schemes lend themselves very naturally to being interpreted as programmed instructions installed in computers. More finesse is needed to view an organism's intentional schemes as residing in the organism per se; for this requires us to ignore Dewey's philosophically appealing suggestion that experience—hence, consciousness and its proprietary schemes—resides in the organism in situ, in the epigenetic system and not in the organism as abstracted from it. But let us assume, for the sake of the argument against the functionalist, that Dewey is wrong and that intended teleological schemes are intraorganically realized.

Going into all the functional details of a system's "data processing" that are needed to establish its credentials as a teleological system might seem to beg the question of its nonequivalence to a system with conscious readiness; for these details will differ between purposive and nonpurposive teleological systems (as my prior analysis of these systems indicates), and some of the differences arguably need not be brought to bear on the question. Fairness to the functionalist may demand that we compare only those functional states that would be roughly parallel in two systems (one purposive, the other not) that, for different ancillary reasons, were judged to be functioning teleologically.

The functional states we might most justly compare are those associated with particular *initial prolepses*—that is, sets of initial proleptic realizations of teleological schemes with a common goal. Let us simplify the discussion by stipulating a certain "single-mindedness" of the systems, that they have no more than one initial prolepsis at any given time. Now a teleological automaton, Autom, must have at least two such schemes in an initial prolepsis (otherwise Autom would lack one sort of many-one plasticity requisite for mechanistic teleology), while a consciously ready being, Anton, need have no more than one; but since there is no proscription against Anton's having more, let us assume (for argumentative parity) that Autom and Anton have the same two schemes in their respective initial prolepses. Assume, too, that Autom and Anton are comparably equipped physically to undertake movements required for the implementation of the schemes.

Both systems are disposed toward the further realization of each of their two schemes—or they could not be said to have each within their initial prolepsis. However, the two systems are not functionally isomorphic, insofar as Autom's selection and final proleptic realization of one scheme is uniquely determined— in existing circumstances—and Anton's is not. (For that matter, whether Anton actually does select one of his two schemes, whether and how he modifies a scheme, etc., are also not uniquely determined; but the argument here depends only on the lack of a functional relation of unique determination, not on the specific character of other relations that may instead obtain.)

Undaunted by the preceding possibility, a cyberneticist might say, "If it is indeterminacy you demand, then it you shall have." Suppose, accordingly, that a randomizing device is incorporated among Autom's otherwise mechanical procedures for selecting a final proleptically realized scheme. Such a device might even make use of lower-order indeterminacy, by having odd or even numbers of Geiger-counter clicks tip the balance in the selection of a final prolepsis, and, so, introduce an element of absolute chance into Autom's data processing. Thus, we may have two systems, Autom and Anton, both exhibiting comparable indeterminacy. But despite all their other possible exact functional parallels, Autom's and Anton's thus shared indeterminacy does not secure the functional isomorphism of their (data-processing model) functioning. Two identically randomized (as above) Automs, fresh off the assembly line, are not even functionally isomorphic to one another, much less to any system with higher-order indeterminacy. Granted even that one might not be able to tell the difference between some (possibly later generation) Automs and the Anton whose functioning they have been devised to simulate, a functional difference would remain between any two of the systems. Even if, *per accidens* of untold improbability, any two such systems should happen to function identically over a definite period of time, those systems would remain, essentially, functionally nonisomorphic—they would not necessarily have to function identically, else they would not have true indeterminacy (of higher or lower order).

At this juncture, the functionalist might take a fall-back position, replacing the here troublesome notion of functional isomorphism with a looser notion of functional equivalence. Doing so would accord with the basic terms of Turing's functionalistic imitation game, but I would not be willing to play along unless this notion were, in use, tight enough to preclude claims of functional equivalence of higher- to lower-order indeterminacy. I would, for example, be prepared to concede that if functioning were characterized according to some data-processing with higher-order indeterminacy model, any Autom functionally equivalent (in the sense of having the "data-processing" competence to pass a Turing-like test of intelligence comparable to Anton's) to Anton would have conscious readiness, too. But this concession is less than it appears, for it is my conjecture that higher-order indeterminacy is superveniently (indeed, uniquely cohibitively) equivalent to conscious readiness; so that Autom's conscious readiness would be not just a purely functional reality but an emergent phenomenon in its own right.

My last concession might be thought something of a cheat, since I have not indicated what sort of *non*-lower-order indeterminacy is here to count as the requisite higher-order variety; but my earlier analysis of flexibility should make an honest claim of that concession: The requisite indeterminacy is non-microlevel-reducible flexibility, is the constitutively irreducible capacity to generate schemes in a final prolepsis that are genuinely novel with respect to those in the corresponding initial prolepsis. This account employs terms that might be thought reserved for characterizations at the level of consciousness, not at the level of its basal physical indeterminacy; but I need hardly remind the functionalist that the terms for functional structures are not thus reserved, given the ontic neutrality of such structures (which are not to be confused with what Russell called structural properties—viz., distinctively physical ones).

I also deny that another, earlier concession to the functionalist was a cheat. To reaffirm that concession in amplified form, I now claim that any system functionally isomorphic to one whose functional structures were realized in conscious readiness would be a system whose functional structures were realized in conscious readiness. Since I deny the very possibility of such isomorphism between systems, my concession may seem less than candid. But I am not simply relying on the wide latitude allowed in making a claim amounting to a null hypothesis. My concession combined with the impossibility thesis is a way of getting at the important (figuratively expressed) point that God, once having conferred "free will" on exact clones, could not reasonably expect them in identical circumstances to behave identically. Besides, my hypothesis is not completely null: There may be said to be exactly one system functionally isomorphic to any given purposive organism—namely, itself. This is not empty rhetoric; it is one statement of the essential individuality of purposive beings.

11.3 Nature and Fate: Some Consequences of the Principle of Sufficient Reason

Contrary to Spinozistic reckoning, the real world, Nature, need not be a rigid arrangement of building blocks, parts, that constitute collectively what James derided as "a block universe." Popper's description of stars as undesigned fusion "machines" (his quotes) suggests that even very simple physical constituents may change their stripes, their elemental natures, when brought together in sufficiently huge numbers. When deterministic laws bring enough simple constituents together, these constituents need not continue to abide by those (or other) deterministic laws. And if the original laws are held in abeyance at the core of that complex system we call a star, they will not necessarily lose all further dominion over subsequent events involving the system's new parts, the products of its old ones' fusions.

Science, philosophers must realize, may change the very character of our appreciation of, our wonder at the stars above us. These stars, symbolic of fate and, for Kant, emblematic of natural necessity, may now be seen, from a scientific vantage point, as rather more than cogs in a closed system of celestial mechanics. Yet philosophers persist in the view that there is such a system at some not too microlevel within ourselves as well as the stars, a system that uniquely determines, from below, reality at every higher macrolevel. Were this view a conclusion carefully drawn from the best available physical theories, it might be hard for the human spirit to bear; but a philosophical apologia could—as some do now—ease the psychological burden, and there could be—as there already is in some quarters—a curious satisfaction in our having discerned with our own fixed capacity for reason this troubling truth. But arguments for the view are not half so strong as convictions about it, and so some further explanation seems called for of why philosophers committed to the use of reason in the service of understanding should be so very sure of something so ripe for cautious, empirically informed skepticism. A full explanation might seem to require psychoanalytic techniques, but I will stop short of them.

My nonpsychoanalytic conjecture about the philosophizing of Spinoza and those who espouse similarly fatalistic doctrines today is that it relies implicitly on some version of that overworked philosophical workhorse, the principle of sufficient reason. Ironically, the very insufficiency of the expressed reasons for such doctrines is what inclines me to this hypothesis. Writ large—as "Nature"— the sufficient reason for the world's changing as it does is also reason for its doing so again should it ever be in the same state. This is reason's shortcut to a deterministic view of the universe, a shortcut that wrongly seems to obviate the need for empirical backing via arguments appealing to established "deterministic" laws and theories and adequate state descriptions of the world. By projecting the principle on the world at large, one receives, for one's armchair efforts, a reflected image of a deterministic universe, bound and determined to do what is in its Nature.

Philosophers have been all too content to rest with some such fatalistic image—witness Russell's causal skeleton. Russell moves from claiming that physics' aim, "consciously or unconsciously, has always been to discover . . . the causal skeleton" to admitting the possibility of surprise "that there should be such a skeleton," to insisting, despite his own earlier avowal of deep indeterminacy, that "physics seems to prove that there is [this skeleton.]"[3] But where is the proof to which we should defer?

Laplace, flush with success at solving every then known problem outstanding in celestial mechanics and Newtonian physics generally, may have had the most reasonable self-earned assurance of any scientific inquirer before or since that the universe is a causally closed system governed by deterministic laws. He stated his conclusion to that effect in terms of what seems a demonic "intelligence" not unlike his own—much more knowledgeable about a total instantaneous state of the universe, but perhaps no more knowing about "all forces acting in nature," the deterministic laws governing it: To this intelligence, were it "sufficiently powerful to subject all the data to analysis, . . . nothing would be uncertain, and both future and past would be present before its eyes."[4] To see either clearly, some heady calculations would be needed, but they would not need to be too heavy, were vision directed toward a causally rather autonomous (so-called closed) subsystem of the universe.

If my analysis of pertinent data is correct, then (never minding the past) only a Laplacian demon who stayed around to confirm all his own predictions could truly be said to know whether determinism is actually true from now until the end of time. Laplace's apparent induction from his successes (on top of Newton's) may justly have inspired confidence that continued confirmations would follow; but with the hindsight available to me from science since, I might ask Monsieur Laplace what would happen when—in accord with his own demon's predictions—worlds, big or small, collide. And I may be fairly confident now, on borrowed scientific authority (as Laplace had been sure then, on his own good authority), that the demon's further predictions—however firmly based on knowledge of deterministic laws—would be incorrect. So much for the causal skeleton; so much for Spinoza's Nature and the rigidly physicalistic world of his intellectual heirs apparent—those supervenience theorists who misuse the concept of supervenience to urge the claim of a microlevel-on-up doctrine of determinism.[5] Whereof physics is unable to speak with full authority, philosophers should not put words in its mouth.

Writ small—as "our nature"—the sufficient reason we behave as we do is also supposedly why we would behave the same way were we ever identically situated. This is reason's shortcut to the claim of instinct. But if our nature is flexible, notwithstanding any native tendencies in particular directions, then the principle of sufficient reason as appurtenant to it may require a different reading. Remember "Buridan's Ass," a problem case named after the philosopher whose theory seemed to suggest, asininely, that an animal poised between two equally appetizing piles of hay set equal distances away would be unable to choose between them and, so, would starve to death in the face of readily available food. Isn't the lesson for us here that hunger is a sufficient reason to turn arbitrarily to the left pile instead of the right; and isn't that same hunger, to the same animal identically situated, also a sufficient reason to turn arbitrarily to the right pile instead of the left? If the principle of sufficient reason cannot be read this way, then that philosophical law is a "Ass." (Natures are what we may know about what we or others will do. There is much we may not know, prior to our own or their actual performances.)

11.4 Toward a Methodological Reconciliation with the Scientific Image

Deterministically inclined philosophers may not be wrong to see as obscurantism the insistence that no mechanistic psychological theory of human beings, or even frogs, is possible—especially not wrong if that insistence is just a mindless refusal even to consider any theory that might be proffered. But, by the same token, it is obscurantism on the opposing side to be unwilling to consider a theory that lacks fully deterministic credentials.

My conjectural commitment to higher-order indeterminacy, to flexibility, has not obliged me to deny that its inception—as a capability—is uniquely determined, nor even to deny that the endogenous results of exercising this capability uniquely determine overt behaviors. I have, moreover, eschewed the idea that this flexibility might be accounted for in terms of lower-level indeterminacy magnified upward and, so, be apparent at the emergent level of consciousness or its basal physical complexity. But this blatant antireductivism does not commit or incline me to further reification of emergent levels as separated worlds that, by a "colossal coincidence,"[6] just happen to be parallel. For that matter, I am no more committed to some (imminently and immanently godlike) one-world hypothesis. Apantheists are not necessarily theists; unbelievers are not believers in spite of themselves. I content myself with the idea that, if there are more worlds than the one I am immersed in, they must be revealed to me by a sober scientist or a rationally persuasive transcendentalist. I need no abiding faith in one world to take this stand.

The physical world is not at some level inherently removed from me; much of that world confronts me rather directly, in experience. The world of physics is my world, too. When I cannot lay hands on its outermost regions or innermost recesses, I am also fully prepared to reach out with other people's well-based scientific theories and, in that indirect way, touch bases with physical reality. I am, in short, a realist who does not naively dismiss the scientific image of reality.

If I have done my job properly, then my conjecture about the special place of higher-order indeterminacy in the teleological functioning required for purposive behaviors should prove well grounded by my philosophical analyses of "ordinary action" and other related concepts belonging to our "manifest image of reality." But scientists are apt to regard my claim as rank speculation that fails to measure up to their standards for the empirical assessment of hypotheses. The rebuke has merit: Philosophers are perhaps too willing nowadays to engage in quasi-scientific hypothesizing, which typically makes mention of causal connections as if they were just evident to philosophers' keen sense of reality. Causality, they tell us, is the cement of the universe; and so they feel entitled to liberal use of the concept of cause in their "naturalized" philosophizing, in conveying a picture, for example, of psychological phenomena in relation to the rest of the world. All the advantages here, as Russell once remarked about something else, of theft over honest toil.

Impatient with the idea of a theory about causal elements of an acceptable theory about such phenomena, scientists may call for a good theory of the latter sort first—on the reasonable supposition that what is needed in the former regard might best be decided not a priori but by what is actually mentioned in an accepted theory, one with firm evidentiary support. "First some theories, only then a theory about them" seems sound advice.

The call for a behavioral theory that lacks fully deterministic credentials seems all the more worrisome, since the best available scientific (microphysical) precedents for such a theory do not meet the implicit requirements (of self-controlled, higher-order indeterminacy) for the theory sought. Yet maybe some semblance of precedence is to be found elsewhere, among ordinary (prescientific?) ways of explaining human behavior. Sometimes referred to as "folk psychology," these modes of explanation should not be summarily dismissed as unscientific. Indeed, if "scientific" psychological theories are reckoned to be inherently mechanistic, then I must register the strong suspicion that they will never explain or predict my doings nearly so well as can already be done by someone who knows me most intimately, without the aid of such a theory.

As a fellow conscious being, I may grasp many of the diverse springs of your acts, the habits and other more *or* less general tendencies that incline you to the particular extent, in the specific direction of your behaviors. My grasp may at times be firmer than your own, and my grasp of my own tendencies may be curiously derivative of my grasp of yours or your grasp of mine.[7*] One person may, for example, know better than another just how much effort the other is likely to expend on a given enterprise: Once, long ago, my father trudged through a snowstorm to buy me a violin—needed at a last minute's notice to avail me of an opportunity to get lessons at school. Dad's stepfather observed knowingly that I would never put the needed effort into the enterprise. Gramps was, as I discovered, correct.

To deny the legitimacy of such a grasp and the remarkably accurate predictions it often affords is to rail against a mode of explanation far less obscurantist than the mechanistic theories that hope in vain to supplant it. Indeed, I suspect that the modicum of explanatory significance had by most psychological theories is derived illegitimately from this unassuming source: Aside from the common pretense that they satisfy the methodological constraint of strict determinism, about all these theories have going for them is an impressive technical vocabulary that really makes sense only as a veiled way of referring to purposive behaviors. Leaving aside pseudoscientific theorizing, we might do well to accept informative, explanatory, often quite reliably predictive descriptions of purposive organisms. What Putnam says[8] of the psychological theories he projects on the horizon of inquiry might also be said of these presently available accountings: They explain what needs explaining at the right level of explanation. And if one is willing to swallow emergence, as Putnam is, why should one strain at the natty new indeterminacy of some emergent levels?

My point is that the explanations ordinarily given of human behavior work in the absence of any distinctively mechanistic assumptions about our conscious superintendence of our acts. My analysis is closely related to such explanations (indeed, they could even be recast perspicuously in terms of it); so that analysis is lent some further empirical credence by those explanations. There is in those explanations a kind of precedent, within the manifest image, for an indeterministic theory of behavior.

But what should such a theory be like? One possible model is Piaget's developmental psychology. His research program, based on careful personal observations of children and on his philosophically sophisticated (though not indisputable) interpretations of those observations, is a welcome relief from the crabbed experimental approach of much contemporary psychology. Among the principal posits of his theory are "action schemata" (which have much in common, might even be identifiable with "teleological schemes") and two processes of human/environmental interaction involving those schemata—namely, "assimilation" (of environmental contingencies to schemata) and "accommodation" (of the schemata to the contingencies). Piaget's observations are made in support of an account, a theory, of how those processes combine to effect a course of cognitive development. The overall picture of that development, its resultant schemata, and the behaviors wrought in accord with them may have been supposed by Piaget himself to be thoroughly deterministic. Nevertheless, that picture need not be so regarded by us: Accommodation, for example, might well be given an indeterminist slant, be interpreted as a genuinely flexible process of adjusting schemata to circumstances.

Although I do not wish to make a blanket endorsement of Piaget's theory,

thus interpreted, it is a further precedent for the sort of theory I would advocate, a clear-cut case of a theory within the "scientific image of reality" (instead
of what was largely a loose assemblage of person-specific generalizations about
human behavior). This precedent goes beyond folk wisdom and might possibly
even be used to derive ordinary explanations of human behavior, and it seems
to allow for indeterminacy, flexibility, at the very core of the determination of
behavior. This might be thought to confer more scientific sense on the idea of
such indeterminacy, but still it would not verify that idea. What would verify it?

Perhaps the very question bespeaks some misunderstanding, some excessively
narrow construal of the relations between evidence and theoretical posits. W. V.
Quine's broader philosophical view of science affords a convincingly less piecemeal understanding of these relations—suggesting that we adjust our theories,
revise their posits, on the basis of evidence that ties whole theories down to earth
only along their peripheries.[9] I propose a major overhaul, a structural rebuilding
of behavioral theory, but it is not obvious that any well-established facts or parts
of theory would be contested in the process. The science of psychology need not
fold its tent after a collapse attendant upon the removal of its rigid framework,
though I do concede that such removal calls for more reconstruction than would,
say, a few holes poked in a patchwork of posits. Yet flexible braces will support
the weight of theory without putting so much stress on the posits or the conceptual
threads that bind them together.

Of course some theorists want as much stress as possible: Methodologically
speaking, how better to test the fabric of theory than to subject it to the rigors of
deterministic supports that serve to radiate the tension of the ties, the evidence,
to all parts of the canvas, the theory? As a heuristic measure, then, we might see
the wisdom of some initial presumption of determinism. But how long must we
go on rending the fabric before it dawns on us that the point of a tent is being lost
in vain pursuit of an unworthy preconception of an ideal tent? We want theories,
not just to satisfy our curiosity about tensile strength of theories, but to afford
the comforts of understanding—as against our present condition, shuddering in
ignorance.

We ourselves are among the principal topics of psychological theory and so
might expect to increase our self-understanding by way of it. But if our most common estimation of ourselves is made in our own "manifest image," then it must
somehow be reconciled with the scientific image that promises a more refined
measure. One extreme position would be to deny the substance of the manifest
image, to regard it as an illusion that does not really afford the comfort of understanding. But I am far less skeptically inclined. I regard the manifest image as
part of the very fabric of any sound psychological theory—a part located (among
other places) right at its periphery, tied securely and most directly down to earth.
I would also contend that only this alleged image can be thus secured, that the
scientific image is never more than indirectly secured. The manifest image can

be said to be tied by manifest actions in the world to the world, though this way of talking will tend to suggest too sharp a dichotomy. The world is made most manifest to us, the point is better put, by our ordinary actions within it. And our scientific theories, in turn, are tied to the world as manifest to us.

Of course some very rarified activities—namely, experimentation and controlled observation—are used to connect our scientific image to reality. And these actions, it is frequently suggested, are very theory-laden enterprises, much more beholden, conceptually, to the scientific than to the manifest image. But before we endorse the further suggestion that we can dispense with this image entirely, we should note that our vulgar manifest image may be indispensable for instructing us in the ways of science. Even so, we might think to dispense with the services of the manifest image after our educations were well advanced. But we would find that it was not so easily cast off. Our confirmatory acts would retain their original association; for actions, as G. E. M. Anscombe has taught us,[10*] may be variously described, and whatever else the scientist may be said to be doing in tying theory to ground, the scientist's actions may correctly be described in terms of the manifest image.

Physics may seem to require relatively little of our manifest image between itself and the world—especially in comparison with psychology, which must depend on that image not only to make appropriate evidentiary connections but also to form the very subject matter of the theories. Psychology must incorporate, conceptually, the fabric of our lives within its theories. This delicate stuff demands a flexible frame, a special indeterminacy; or we rend the fabric, stitch some coarse material to what remains, and pretend that psychology need be no more indebted to naive, prescientific conceptions than physics is. This pretense is radically mistaken not just in its suggestion that we forgo the earlier conceptions of ourselves but in its antecedent supposition that physics is so much at odds with the manifest image—when in truth physical categories are, by and large, either already contained in or quite easy to reconcile with that image, provided no attempt is made to reconstitute it entirely of those refined categories.[11*] The suggestion that we forgo the manifest image of ourselves in favor of one newly reconstituted of categories far less refined than those of physics and philosophically more naive than those of the manifest image has little to recommend it— especially since no distinctively deterministic theory of the sort that is supposed to warrant that suggestion has ever amounted to much.

11.5 Emergent Indeterminacy and Unified Science: Some Neurological Speculations

I can work up much more enthusiasm for Skinnerian behaviorism than might be supposed on the basis of my preceding remarks. As Skinner observes, "It is not the behaviorist, . . . but the cognitive psychologist, with his computer-model of the mind, who represents man as a machine." [12] Skinner's behaviorism might not unfairly be described as a probabilistic approach to the study of behavior—and as such, there is nothing inherently mechanistic about it. His approach requires no concurrent support from physiological findings, though in their absence he is not entitled to judge whether the probability of a particular operant response's being emitted under the control of a stimulus is a matter of randomized but determinate chance or genuine indeterminacy.

My own more philosophical inquiry has led me to aver that the mental flexibility/physical higher-order indeterminacy duals are at the core of the determination of purposive behavior—that their being there is, in other words, a conceptual requirement for genuinely purposive behavior. I could stick to this story even if a scientific account to support it were never forthcoming, even if a deterministic theory of all behavior were to be firmly established; for, just possibly, purposiveness itself may be an illusion.

Or, in the event of a Piaget-style theory on firm empirical footing but with no deterministic credentials, I might be supposed to garner support for the extra idea that my conceptual requirement is actually met, so that some behaviors really are purposive. But surely more than a psychological theory based on observations (however insightful) of overt behavior would be needed to verify the physical side of the story, needed to support the specific idea of higher-order indeterminacy.

Given the nature of the beast, perhaps the obvious scientific discipline to address this need is neurophysiology. But in proposing this candidate, I am not reverting to an inveterate reductivism; I am not supposing that psychology must ultimately be explained in exclusively physiological terms. Still, perhaps I am inclining toward some other, nonreductivistic, "unity of science" ideal. The tent metaphor may be used to conjure the idea of stitching together all our theories into one grand integument. The point of doing so would be not merely to understand one science-specific range of phenomena apart from another but to comprehend all phenomena under one roof—to understand how everything "hangs together." This particular philosophical spin on the idea of unifying sciences suggests a bad tendency to prejudge the nature of reality, since our valid theories may, for all we now know, only hang loosely together as, for example, a cluster of smaller tents, with different frames and different ties to reality. But on a less grandiose scale, the idea has merit—some things do seem to hang together already (despite their different natures and the apparent autonomy of the scientific disciplines that study them); so we may well wonder what type of tie may bind.

To fathom what may be the basis for the unity of my alleged two aspects would require deep metaphysical inquiries beyond the scope of this discussion, so suffice it to say that such unity is more than parallelism but less than identity. Dual aspect theories are often criticized for their obscurity, as if in deference to the very terms of Descartes's suggestion that his duals, mind and body, were clearly and distinctly perceived to lack unity. But assessed by more recent standards of clarity (or transparency of implications), the (not yet double-aspect) claim of mutual supervenience should escape such criticism; for while contemporary supervenience theorists may not have compelling arguments for their physicalistic/deterministic views, they have framed clear concepts for use in stating those views (or assumptions) and my denials of some of them.

Given a desire to understand, from a scientific perspective, the real character of the supervenience-theoretic tie that binds conscious flexibility to physical higher-order indeterminacy and both to other natural phenomena, we might reasonably turn to neurophysiology for help. But since no extant theory quite satisfies this desire, we might have to content ourselves with wondering what might satisfy it and, on the basis of apposite neurophysiological findings and theorizings, speculating accordingly.

On the face of it, there is some puzzle about how an organ such as the brain, whose component neurons seem so determinate in their functioning, might possibly give rise to higher-order indeterminacy. Yet even if we reject the idea of a higher-level (though smaller-scale) parallel to stars as fusion machines, of novel schemes "cooked up" in a neurological cauldron, we might still suppose that the very complexity of the brain has led to something new, to emergent indeterminacy.

We might think to argue from the less-than-perfect reliability (for their biologic purposes) of individual neurons to some net indeterminacy of brain functioning, but unreliability seems here confused with indeterminacy and hardly seems the boon the latter could be. The point is *not* to achieve unreliability on high. Achieved reliability is surely more to the point of any system as such; in fact, reliability is largely what it means to be a system and not just an aggregate of randomly interactive phenomena. According to Ryle, "Paradoxical though it may seem, we have to look [not to inanimate groupings but] to living organisms for examples in Nature of self-maintaining, routine-observing systems." [13] And this constancy may be said to result from an organism's profuse many-one plasticities, which allow for more reliable outcomes—including very regular functioning—than might otherwise be had in the naturally chancy situations nature provides.

John von Neumann, who depicts living organisms as "natural automata," stresses that the means whereby automata meet the problem of reliability of functioning are not so much "precautions against failure" as "palliatives" in the face of failures that are a normal part of real functioning (as opposed to the ideal functioning of axiomatic systems that are not substantively realized). I agree that humans qua organisms are automatistic with respect to much of their biological functioning and that much of the success of that functioning depends on their inherent means of establishing reliability, but I do not see any point in attributing the nonautomatistic, indeterminately flexible behavioral functioning of human beings to residual unreliability. For far from its being a further failure, that functioning is a superadded success promoter, a reliability improving measure that facilitates successful coping with the environment.

Might this superaddition be a direct consequence of further increases in a system's organized complexity, of much more convoluted but still reliably functioning neurophysiological structures? Judging from axiomatized idealizations, one might quickly answer no, since axiomatic systems seem not to allow for the derivation of any pertinently novel item. But von Neumann suggests, on the basis of some of A. M. Turing's formal results, that complexity of an idealized system may reach a level beyond which "the description of what a mechanism is doing [—that is, its rule-governed 'behavior,' say, in churning out halting or nonhalting strings of symbols—becomes] more involved than the description of the mechanism, in the sense that it requires new and abstract primitive terms, namely higher types." K. Gödel, whose conjectural reconstruction of von Neumann's suggestion I have just quoted, further clarifies the idea by noting, apropos of a universal Turing machine (which could effectively mimic the computations of any automata), that "it might be said that the complete description of its behavior is infinite because, in view of the non-existence of a decision procedure predicting its behavior [say, whether it will halt], the complete description could be given only by an enumeration of all instances." [14] And von Neumann sees some "harmony" between this point and the further curiosity that "there is a minimum number of parts below which complication is degenerative, in the sense that if one automaton makes another the second is less complex than the first, but above which it is possible for an automaton to construct other automata of equal or higher complexity." [15]

These observations about idealized (or formalized) automata may augur well for the idea that sheer complication can engender some interestingly novel features, but is any more specific encouragement thereby given that higher-order indeterminacy will be among those features? It is perhaps misleadingly hope inspiring that, as the first observation has it, one cannot predict the halting of the Turing machine, for there is still a sort of prediction makeable, in principle, by direct observation of another, functionally isomorphic automaton; and this sort of predictability confutes the indeterminacy hypothesis.

But von Neumann's second observation may still engender hope. Beginning with an avowedly intuitive but vague idea of "complication"—as "effectivity," "the potentiality to do things" [16]—von Neumann poses a dilemma about automata that "synthesize other automata from elementary parts": Organisms, in the course of their phylogeny, sometimes become more complicated. That would seem to suggest that these natural automata "have the ability to produce something more complicated than themselves." [17] Thus, synthesis here seems "explosive" rather than degenerative. And yet, on the other horn, reflection on artificial automata seems to suggest that automatistic synthesis of automata is inevitably degenerative—given "that a machine tool is more complicated than the elements which can be made with it, and that, generally speaking, an automaton A, which can make an automaton B, must contain a complete description of B and also rules on how to behave while effecting the synthesis." [18*]

Von Neumann breaks the second horn of his dilemma by constructing idealized automata that produce other either equally or more complex versions of themselves from available component parts. The first trick, of equal (in fact, physically identical) complexity, is achieved by dividing reproductive labors among several component automata: 1) "a universal machine tool A which, when furnished with $Phi(X)$ [an axiomatic description of any automaton X], will . . . take it and gradually consume it, at the same time building up the automaton X from [available] parts"; 2) "an automaton B which "consumes $[Phi(X)]$ and produces two copies of [it]"; 3) "The automaton C [which] dominates both A and B, actuating them alternately [as follows:] i) C has B make two copies of $Phi(A+B+C)$, ii) C has A consume one copy of $Phi(A+B+C)$, while constructing $A+B+C$, iii) C ties the remaining copy of $Phi(A+B+C)$ to the newly constructed $A+B+C$, and detaches them from itself as conjoined to A and B, i.e., from the old $A+B+C$." *Ecce* auto-reproduction.

The next trick, synthesis of more complex counterpart automata, is managed by letting X in $Phi(X)$ be $A+B+C+D$ (instead of being just $A+B+C$, as above), where D is any automaton. Then steps i–iii, still taken by $A+B+C$, yield the automaton $A+B+C+D$, which has greater complication than $A+B+C$ itself.

This procedure might seem to allow for flexibility on the part of any automaton whose complication equals or exceeds that of $A+B+C$; for this particular automaton would seem to have the ability to produce more complex, hence novel, versions of itself. If so, then complexity per se might be thought on a priori formal grounds to be capable of yielding emergent flexibility of a strictly deterministic nature. Complexity would not have engendered any higher-order indeterminacy, but this essay would no longer stand in need of any: Determinism would have been shown to allow for genuine flexibility, hence, consciousness.

But there is ample reason to conclude that $A+B+C$ (or S) still exhibits no real flexibility, for S's initial prolepsis must (if S is to have the determinate ability to synthesize its more complex counterpart) contain the very teleological scheme for the production of the other automaton $A+B+C+D$ (or S') which is finally proleptically realized. Suppose, though, that S had been poised for exact self-replication, toward which goal S had many-one plasticity, but that some mutation has modified S's goal, by changing $Phi(A+B+C)$ to $Phi(A+B+C+D)$. Two possibilities ensue: Either S's reproductive plasticity survives, invariant save for a new goal, or it does not. If the former, then a new initial prolepsis (with the final scheme) effectively supplants the old one, which lacked that scheme; if the latter, then the final prolepsis (with that scheme) is also an initial prolepsis. In all cases, then, the scheme is included in an initial prolepsis, so S exhibits no flexibility.

It is important to note that goal shiftings are not plausibly regarded as organic abilities of a system lacking intentions and caused—by factors external to its teleological plasticities—to make those shifts. Thus, the first horn of von Neumann's dilemma is miscast: Our lowly phylogenetic ancestors, for example, had no ability as such to produce us, and though one organism may have a specific ability to produce an offspring more complex than itself, it has that ability only after mutation "reprograms" it. Nonetheless, von Neumann's idealized support for the first horn, properly recast, does contribute to an understanding of how evolution might, on the basis of reproductive reliability measures, manage its "trick" of preventing or even reversing quasi-thermodynamic/information-theoretic tendencies toward degeneration of complication.

But highly complex reliability-ensuring procedures are convergent plasticities and so cannot in and of themselves be supposed to provide flexibility (or, then, consciousness). Indeed, on the assumption that those procedures could yield flexibility if their complication were vast, one might expect bodily injury repair procedures involving the replication of cells to have as much complication as simple actions, if not more, and, so, to be flexible; yet we consciously perform actions but never consciously make such organic repairs.

Of course even though complex reliability ensuring procedures do not formally secure flexibility, it is possible that as substantively realized in grey matter, they as a matter of fact do. Moreover, whether or not directly achieved by means of convergent plasticity, reliability of production is only to be expected in the morphogenesis of consciousness—otherwise, evolution could not "count on" its adaptive benefits. Unless consciousness is reliably wrought, it is not reliably available for its purposes.

Without any real sense of the relevant numbers or even of the right metric for use in gauging complication, I am inclined to think that the relatively simple fertilized human ovum in situ is extremely explosive with respect to its crowning morphogenetic achievement—the synthesis of that most complex organ, the human brain. Even granting that a fertilized ovum doth not a whole brain make—that interaction among successor cells as grouped may influence the character of those cells' subsequent synthesizing of still more cells, and so on—the need for explosion seems palpable: The original cell, on grounds of relative complexity alone, can hardly be said to contain a blueprint of, to be isomorphic to, the final product, yet that product seems automatistically wrought by that cell and its successively synthesizing successors, by generation upon generation of variously dissimilar replicants.

Viewed phylogenetically, explosion might be brought about by advantageous mutations, by changes in the genetic program that fortuitously exploit previously attained levels of neurological organization and carry it to still higher levels. For all I know, such changes might be minor in comparison with their consequences, but this just suggests that explosiveness could be explained, not that it is unnecessary.

One striking further feature of neuromorphogenesis is the curiously prehensile look of loose nerve fibers that, during morphogenesis, flail about until they make contact with and affix to other nerves—making for associations that may be long or short term, ending with the demise of one or more associates. Of course this inchoate prehensiveness is merely imperfect and not the key to the conscious flexibility attendant upon subsequent neurological organization; for prehensile flailing ends when associations are made. Yet fibers flail 'til flexible heads prevail; so might not this seemingly aimless flailing be evidence of each neuron's ever-so-rudimentary but more general prehensiveness, a degree of flexibility one-X-illionth of that required for conscious readiness in furtherance of molar-level behavior? Such prehensiveness might first express itself in flailing but then secrete other modest contributions to higher-order indeterminacy. Perhaps too much is made of each neuron's discrete discharges, its digital/computational contribution to an overall "flow of information"; perhaps each neuron has something of a more analogic (or nondigital) nature to offer as well. Let us not be too quick to dismiss analog responses, continuous properties, of neurons as idle periods of fatigue that are incidental to normal overall neurological functioning. Might they not be periods of incomplete cogitation, of inchoate rumination, instead?

11.6 Prospects for Further Research on Flexible Functioning

Whether conscious readiness is a wholly unprecedented emergent making its first appearance on high, at levels of extreme neurological complication, or whether that appearance is foreshadowed down low, in the not quite prehensile qualities of individual neurons, is not for me here to decide; but both possibilities (and others besides) are, I contend, not unreasonable speculations, given the current state of scientific knowledge.

What more might now be said, and what more might future empirical inquiry be expected to say about the place of consciousness in the neurological-world order? Somewhere between stimulating input and responsive output lurks conscious readiness and its schemes. The brain does seem the likely chief locus of this mediating consciousness, even if input and output, which have their own conscious aspects, do reside in a larger habitat. But where exactly might the center(s) of conscious control be found within the maze of neurophysiological structures? An analogous question, say, about the exact location of a modern-day computer's "random access memory" seems difficult enough to answer, though an electronics expert could be expected to tell us just how RAM pulses and circulates through the hardware; but the question about our brains seems disanalogously more difficult, since we are looking for a genuinely unknown source and are lacking a coherent method to trace it. We might suppose that someday we shall find that source; but we have no reason to be confident, and Wittgenstein, for one, saw no call to be:

No supposition seems to me more natural than that there is no process in the brain correlated with associating or with thinking; so that it would be impossible to read off thought-processes from brain-processes. I mean this: if I talk or write there is, I assume, a system of impulses going out from my brain and correlated with my spoken or written thoughts. But why should the *system* continue further in the direction of the center? Why should this order not proceed, so to speak, out of chaos? The case would be like the following—certain kinds of plants multiply by seed, so that a seed always produces a plant of the same kind as that from which it was produced—but *nothing* in the seed corresponds to the plant which comes from it; so that it is impossible to infer the properties or structure of the plant from those of the seed that comes out of it—this can only be done from the *history* of the seed.[19]

Why, indeed, should there be a central system per se? Wittgenstein employs a rather murky conception of potentiality, and given our current biochemical understanding of plants, his analogy may strike many people as quaint and curious biological lore: The DNA molecule, they would say, corresponds in the seed to the plant produced. A modern epigeneticist, though, might take the wisdom of Wittgenstein's remarks to be based on the idea that potentialities are less inside an organism than within the larger epigenetic system, and Wittgenstein's appeal to the importance of the seed's history could seem to support this reading. According to this view, then, conscious readiness, for example, might be placed somewhere beyond as well as within the confines of the CNS.

But Wittgenstein's own prominent mention of chaos would seem to suggest, in addition, a more metaphysical supposal, one drawing on a common if not natural human mode of cosmological explanation—on the myths or the theories of creation as order from chaos. Given this supposal, even if a more extensive domain than the CNS were determined to be the true habitat of consciousness, its location within that habitat might prove refractory to all scientific or philosophic inquiries. We might never make a conclusive identification of the source within the chaos.

This not unnatural supposition can be made to fit two major constraints plausibly imposed on conscious readiness. First, just as a seed always produces a plant of the same kind, ontogenesis reliably produces the kind of neurological chaos upon which conscious readiness supervenes. Second, and no doubt closer to the intended point of the seed analogy, certain specific chaotic states of the CNS reliably produce impulses correlated with spoken or written thoughts and with other overt actions.[20*] These constraints would place conscious readiness, fore and aft (i.e., at its inception and at its final prolepsis), squarely in the causal nexus but still allow this consciousness to bulge flexibly during the interregnum.

Further details may be forthcoming from science, but if Wittgenstein is right we cannot count on science to give so much as the complete physical side of the story. To begin to examine the prospects on the psychological side, let us consider how far the story on that side has already taken us: The story I told was intended to be anti-instinctive in the extreme, but I might not have had to go as far as I did—might not have gone all the way to indeterminate prehensiveness—in order to counter the claims of instinct. Thus, an epigenetic viewpoint that rejects strict genetic unfolding generally and places behavioral potentials extraorganically wins the day against most serious instinct theories. However, I also wanted firm possession of my acts, moral responsibility for them, and an epigenetic view tends to erode personal autonomy.

Now even on the assumption that my anti-instinctive, provoluntaristic stand is basically correct, I am inclined to think that additional psychological inquiries might serve to approximate more closely the exact truth about instinctivity. And, given some willingness on the part of instinct theorists to affirm the reality of flexibility, I am quite prepared to entertain their conjectures about the existence of instinctive elements in behavior.

I have sought more to defang than to destroy my black beast, instinct. But now that I have robbed it of its fatal, fateful teeth, I would be happy to admit it in my living quarters—provided someone manages to catch hold of it. I am even prone to think that some such beasts reside in me, that many of my prehensile tendencies toward ends vital for me, for my species may have instinctive elements. Otherwise, I am at something of a loss to understand why I should grab for certain things so avidly. Yet I do not suppose I can simply intuit the difference between Nature's imperatives and my own self-dictates—any more than I can introspect habits in my conscious determination of acts.

The least instinctive hypothesis about my own intentional behaviors is that they are sophisticated successors to an original prehensiveness devoid of environmental information but equipped with mechanical means and the ability to learn to exploit them. If a more instinct-laden view of actions is to prevail, then it must, while conceding our actions' flexibility (if not on philosophical then on practical grounds), use scientific measures to snare our inner beastliness. Having shown how instincts, if flexible enough to count as more than knee-jerk responses, might wriggle through the most rigorous trap ever set for them, the deprivation experiment, I am inclined to think a less rigid device would be worth a try. My suggestion is that we might follow Piaget's programmatic lead but attempt to challenge his theory by confronting it with the least instinctive hypothesis.

What I have in mind would be a reexamination of Piaget's story of early development, attempting to reinterpret it as a progression from the most minimally nativistic starting point philosophically conceivable. Recalcitrant facts, data incompatible with any such reinterpretation, would then be seen as evidence of a vigorous beast, seized at last.[21*] Some may see such a procedure as more philosophic than scientific, and in truth it would mark something of a return to a time when philosophy and psychology had not yet split; but let us not be too ready to shun the wisdom of the past or to shy away from possibly fruitful collaboration in the future. My own sentiments might be put by aping one of Plato's maxims:

The evils of theories of intended behavior will not end until philosophers become psychologists or psychologists become philosophers.

11.7 Beyond the Bounds: Sociobiological Constraints on Being Human

The way of truth about behavior requires joint practitioners of both disciplines, philosophy and psychology, to take a biologic turn, too. But there is no special magic in multidisciplinary efforts, and there is the great danger that people who wear too many hats will end up talking through them. Consider the case of Edward O. Wilson, the self-styled "sociobiologist" who too freely extrapolates from mixed empirical findings to conclusions favoring what some of the sharpest critics of his movement call "biological determinism."

As if lamenting that man is born in bondage but everywhere behaves too freely, Wilson opines that Nature must keep us on a rather long leash. He warns us off social arrangements that might test the limits of our ordained compass, hinting that the biological price could prove too high. One almost feels a collar tighten around one's neck as one contemplates making a radical break with the past, however distasteful that past might be: Even war, he ventures to say, might be just the stress and strife we need to preserve our humanity—our good qualities coupled with our more destructive ones. Deep emotional pleasures and greater biological efficiency are touted by Wilson as likely benefits of social circumstances that suit our natures, as favors conferred for staying within the circumference permitted us by that stern dominatrix Mother Nature. This man wants discipline.

Critics of Wilson have been accused of confusing the sober scientific issues with political innuendo, so I will not. But I will urge that multidisciplinarians be held to the highest standards of their several callings. Too often a reviewer will suggest that although a book is weak if not inept in the reviewer's own area of expertise, the work does have redeemingly insightful things to say about other fields, fields concerning which the reviewer admits to being ignorant. Wilson is admitted by some of his staunchest critics to do excellent work within his rightful academic field, entomology, which in fact overlaps with theirs. But we should think twice about the substance of everything urged by a person who reasons injudiciously in any area on the supposed foundation of other careful research. Authority exceeded is authority at risk.

Slavery figures prominently in Wilson's sociobiological worldview, and the Sociobiology Study Group suggests that in remarking on the "slavemaking" of ants, Wilson seems to project a human institution upon insectan society and then to see it as a mirror of our own. Wilson's reply is to ask, rhetorically, of the group, "Do they wish to expunge communication, *dominance,* monogamy and parental care from the vocabulary of zoology?" [22] (emphasis mine) Well, if his basis for the use of these terms is no more compelling than that for regarding as slavers those ants that bring immature ants of other species back to their own nests, where the "captured" ants later perform housekeeping tasks, then Wilson's use of all those terms merits critical scrutiny. The group notes that "domestication" fits the ant case at least as well as "slavemaking" does [23]; the case could even suggest "adoption."

But while Wilson is quick to find slavery among ants, he is in less of a hurry to condemn the practice among members of his own species. Wilson, in the austere guise of a disinterested scientific observer, sees the "slave society of Jamaica"— replete with debauched clerics; rampant promiscuity among slaves and their demanding masters; scant education, justice, or cultural "refinements"—as reassurance that "deviations" from societal norms may be extreme yet pose no real biologic threat: He notes that "for nearly two centuries . . . people multiplied while the economy flourished." [24] Indeed, so "flexible" is the "human biogram," that little seems to alarm Wilson except efforts at "social control" to curb baser instincts.

By conceding the possible existence of nativistic elements in flexible prehensions, I may appear closer to a Wilsonian viewpoint than I would care to confess. But my view actually offers no aid or comfort to Wilson. He may really be closer to Emerson, who also understands our Fate, figuratively, as a subtle yet inescapable form of bondage: "When the gods in Norse heaven were unable to bind the Fenris Wolf with steel or with weight of mountains,—the one he snapped and the other he spurned with his heel,—they put round his foot a limp band softer than silk or cobweb, and this held him; the more he spurned it the stiffer it grew." [25] My concession of possible instinctive elements is a far cry from any suggestion that instincts might be deceptively limp bands that allow a degree of movement but tether more effectively than steel. Such bonds have more definite bounds, beyond which further struggle is useless—for once the tethers pull tight, they yield no more gains, only pain.

If Nature had sought slaves to do its bidding, it would, on my view, have played a grand trick on itself; for by granting us enough flexibility to carry out its many schemes Nature would have given us the means to disobey—to bend the rules, to go beyond its dictates, even to strike out in new directions. And given our ability to profit from our past, to use its lessons, to build on it, we pay no inevitable price in loss of humanity, satisfactions, or efficiency—we can and do lose, sometimes everything; but, on a more positive note, there are no clear prior limits on how much we may stand to gain.

On what grounds have sociobiological constraints been affirmed? Have we ever run up against any and failed to transcend them? Or is Wilson just extrapolating from some fairly widespread human traits to grandiose claims about human nature and dire warnings against challenging it (read "him")? "Human beings are absurdly easy to indoctrinate—they seek it." [26] Is Wilson simply noting this rule or applying it? Is the application perfectly general? *Nosce te ipsum.*

Other social species—ants and termites, various carnivores and primates lower

than we—are said by Wilson to have less freedom in consequence of more competition, because they "tend to be tightly packed in the ecosystem with little room
for experimentation or play." [27] But while we might well behave more predictably
in an environment with little latitude in which to exercise our freedom, I rather
doubt that ants freed from harsh ecological-economic realities would prove much
more flexible than they are now. In any case, Wilson's intent is to explain why
we are more flexible than his principal research subjects and other social animals,
but his remarks seem so oblique as to reverse the proper order of explanation:
More plausibly, it is because we are so flexible, so downright crafty, that our ecosystem lacks any serious animal competitors besides our fellow Homo sapiens.
And insofar as Wilson means to wax metaphoric, to suggest that Nature allows
more sportive variation to thrive when interspecific selection pressures are weak,
his claim still ignores the "edge" afforded any rival even slightly more flexibly
adaptive, more intelligent, than the competition.

Given a proper measure of reasonableness, it would be unreasonable to expect
certainty from empirical inquiries, to dismiss theoretical conclusions drawn from
statements of evidence just because the former are not logically deducible from
the latter. But in going beyond the evidence that suggests them, good scientific
conjectures should, as Wilson's sociobiological ones do not, fit integrally in a coherent pattern of explanation that might, in a clearly conceivable way, be securely
tied to reality. To suppose otherwise is to give up on the scientific search for truth,
not just on an ill-conceived philosophic quest for certainty.

Wilson's is just one in a long line of misguided attempts to affirm instincts,
social or not, on the basis of a synthesis of available information about animal
behaviors. Wilson may properly judge his own efforts as more sophisticated than,
say, Herbert Spencer's, but only because he has more and better information, not
because he clearly possesses a closer approximation of the truth about instinct.
He cloaks his claims about biological constraints on behavior in cautious qualifications to the effect that we do not know but what there are such constraints. This
is recognizable as skepticism, to which I respond in kind: We do not know but
what genuinely flexible prehension is a key ingredient of every behavior presumed
instinctive by Wilson.

11.8 Flexibility as Fate

Scientific inquiry should aim to satisfy the highest curiosity—a desire to get to
the bottom of things, to figure them out in a general way that establishes the truth
about them. Philosophers may assist this noble enterprise, even pursue it along
with scientists; but there is a point at which philosophers must hold back (or
change hats): Picture scientific conjectures as finger extensions, as tools designed
to extend the range of possible apprehensions of reality. To grasp something by

means of such devices, at their farthest reaches, one needs a thumb extender, too—that is, observations and experiments cut to the opposable measure of a theory, capable of being coordinated with its individual conjectures. Philosophers may not engage in empirical labors with these particular implements, but not for want of due appreciation of such efforts. Philosophers are simply in another line of work: They attempt further comprehension, by cool reason, of things as already apprehended by other means, including fantasies. Theories as supported by evidence are fit topics for possible comprehension by an informed understanding, with the sole assistance of reasoning—the facts are in, supposedly; and further philosophic inquiries about whether they are in, really, do not inevitably require more empirical thumbwork. Philosophic conjectures may improve our reach in various ways, but these conjectures do not need, for successful grasping, extended empiric opposables.

Flexibility, as I have depicted it, is off the map of the nature-nurture controversy, whereas the sort of flexibility commonly ascribed to instincts (by Wilson and others) would seem to be a set range on the axis between nature and nurture. My philosophic survey of the conceptual terrain of the claims for instinctivity persuades me that they are most fully opposed by flexibility as I have portrayed it, so I have sought to dislodge the question of instincts from its customary place in the nature-nurture debate. But dialectical maneuvers have a curious way of resurfacing in new arenas, and so it is that the Emersonian image of Fate might be thought applicable to my notion of flexibility.

Such flexibility was said to be a certain active prehensiveness, requisite for purposive behavior and capable of underplaying or overplaying any preset plan—all of which suggests a wide range of possible performances. But quite apart from whether there are any definite limits to that range, there is another sort of boundary beyond which flexibility qua consciousness cannot stray: Conscious flexibility is fated not to lose its prehensiveness.

This may appear to be the "mere" conceptual point that *consciousness* implies *prehensiveness,* but it is really a world-necessity known a priori—once we have been around awhile. Philosophy, unlike science, may properly be a quest for necessity, so the point seems more philosophic than scientific. Yet there is even a sort of substantive prediction possible on the basis of this synthetic a priori necessity: If you try to rid your mind of things not by active banishment but by letting go, by releasing your grasp of them, you will, proportionate to your success, also attenuate your consciousness. Such largely self-induced alterations, diminutions, of consciousness as transcendental meditation (or relaxation response) and hypnosis, both of which rely on increased passivity to induce their changes, attest to the correctness of the general principle. Another test would be at bedtime to let one's mind wander freely over thoughts and images, forcing none of them in any particular direction, seizing on none. To sleep, perchance to dream is, I predict, the likely outcome of thus relinquishing control of one's mental economy. Philosophy might here be said to have become psychology or vice versa, but I prefer to regard this as a genuinely interdisciplinary result.

This result suggests, further, that resignation is not true freedom: The mind is active, so free to flex; or it is not. Emerson's figure suggests another sort of futile struggle against Fate, and had he restricted the scope of his suggestion to Circumstance, then I would agree with him: Some contingencies render struggle useless. But pace Emerson, the internal limitations of our flexibility impose no analogously definite (though inestimable) outer bounds.

To make this point plainer, I should amplify my account of purposive behavior in one crucial respect. As it stands, my account seems dangerously close to an ideomotor theory that allows for some free play of ideas, intentions, followed by an utterly mechanical implementation of them in their final form. The locus classicus for such a view is William James's theory of volition.[28] According to James, the ideas that pop into our heads—as uniquely determined by experience and associative laws—are, singly, irresistible springs to action (or reaction). Only a contrary idea, itself brought about without our real complicity, can resist the dominance of a single idea. The stronger idea, the one closest to the focus of our attention, is what then holds sway. But to some extent, the voluntary effort of attention, in thinking, is capable of shifting our focus so that the initially more marginal idea prevails over us. Voluntarily attending to one idea rather than another is, for James, the source of elastic behavior—or free action.

But this theory, while it does not view our actions as robotic, would seem to regard them as overly bionic. What seems ignored is the role played in our overt actions by our physical exertions. My account is open to the same complaint, for though I have indicated that intentions are partwise resolved upon, each part at the point of its final resolution (or prolepsis) would seem, on my account, to allow for little more than purely mechanical implementation. Yet what of the manifest role of variable effort, which in the very process of implementation further modifies, bends, our schemes?

One way to answer the charge of having denied the manifest reality of this role would be to note that efforts can modify only not-yet-implemented parts of a scheme and to suggest that the lived experience of effort's pain is but the sensory feedback from, say, recalcitrant circumstances that did not permit the full realization of the scheme part whose implementation had just been attempted. But the feedback suggestion is questionable, since undertakings seem more immediately arduous than it would have them be; and while, for example, pressing one's finger against a sharp object may cause pain, this is not the pain of the effort but only incidental to it. Might effort's pain be feedback from inner sources of opposition, from bodily circumstance—muscular contractions nudging nerves? Even this seems curiously after the fact of the exertion that is painfully onerous in its own right.

A comparison with purely mental effort in covert problem solving seems telling: Our conscious effort to grasp a solution to a problem can be painful, yet no meningeal squeezing of neurons accounts for our agony. Purposive efforts are struggles to realize, overtly or covertly, teleological schemes. Conscious effort, a phase of conscious readiness, is the means whereby those schemes are bent, whether only in thought or also in the very process of physically implementing them. If one is of an odd mind to press one's finger against a sharp object, then the effort is not the pain but is, partly, one's willful persistence in face of it. Physically taxing muscular exertions are not as liable to damage the body by way of the environment, but they may require near masochistic "will power" in the face of the endogenous agony to which they are heir. Mentally taxing intellectual efforts are comparable in this regard, though they are hardly the basic model on which to understand the physical ones, which are still more elemental. We are our bodies first and last, and our abstract intellections are not the "simple natures" Descartes claimed them to be. Effort in the course of overt actions is the primary means whereby we bend our schemes to circumstance. Deliberation may give us added flexibility, but once a scheme is selected thereby, that scheme almost always requires further resolution via conscious efforts to make it work.

The limits of our efforts are not as definite as Emerson implied. The "bands" he spoke of are not so much limp as elastic, and when we pull against them our efforts are less obviously futile: We stretch these bands, and they—perhaps more transilient than resilient—thereby acquire still greater stretchability in the direction of our stretching. The danger of a breaking point seems highly overrated, given our capacity for growth well past morphological maturity—broadly speaking, educational growth, whose biologic basis might well be some unintended physiological tendencies to reinforce the physical reorganizations attendant upon our willful exercises of flexibility. But even if there is a point of diminishing returns for stretching in any one direction, a point at which our efforts are insufficiently repaid, we may branch out in (some from among uncountably many) new directions when we reach that point, and so persist in pulling beyond our prior limits, transcending our transient natures.

Notes and Index

Notes

Asterisks after note numbers in the text indicate that the notes are not or are not merely citations.

Chapter 1

1. Jean Piaget, *The Origins of Intelligence in Children*, trans. Margaret Cook (New York: International Universities Press, 1952), p. 24.

2. Ibid.

3. John Locke, *An Essay Concerning Human Understanding* (New York: Dover Publications, 1959), p. 39.

4. Charles Darwin, "Posthumous Essay on Instinct," in G. J. Romanes, *Mental Evolution in Animals* (New York: D. Appleton and Co., 1984), p. 379. Romanes says this material was "written for the 'Origin of Species,' but afterward suppressed for the sake of condensation" (p. 355).

5. D. O. Hebb, "Heredity and Environment in Behavior," *British Journal of Animal Behavior* 1 (1953) 1:43–47. Reprinted in D. E. Dulany, Jr., et al. (eds.), *Contributions to Modern Psychology*, 2d ed. (New York: Oxford University Press, 1963), p. 7.

6. Ibid., p. 4.

7. Ibid., p. 5.

8. B. F. Skinner, *About Behaviorism* (New York: Vintage Books, 1976), p. 58.

9. Ernest Nagel, *The Structure of Science* (London: Routledge and Kegan Paul, 1961), p. 333. Nagel expresses serious misgivings, though, about the notion that such emissions are "absolutely chance" occurrences.

10. Hebb, "Heredity and Environment," p. 6.

11. Ibid., p. 4.

12. Ibid., p. 7.

13. Ibid., p. 4.

14. Ibid.

15. Konrad Lorenz, *Evolution and Modification of Behavior* (Chicago: University of Chicago Press, 1965), passim.

16. Ibid., p. 83. Variants of the technique that involve social isolation are sometimes called "Kasper Hauser experiments," after the famous "Foundling of Nuremberg." Cf. W. H. Thorpe, *Learning and Instinct in Animals* (London: Methuen, 1956), passim.

17. Lorenz, *Evolution and Modification*, p. 85.

18. Ibid., p. 96.

19. Ibid.

20. Ibid., p. 11.

21. Ibid., p. 90.

22. Ibid., p. 108.

23. Ibid., p. 23.

24. Ibid., p. 104.

25. Daniel S. Lehrman, "Semantic and Conceptual Issues in the Nature-Nurture Problem," in L. R. Aronson et al. (eds.), *Development and Evolution of Behavior: Essays in Memory of T. C. Schneirla* (San Francisco: W. H. Freeman and Co., 1970), p. 36.

26. Ibid., p. 30. After T. C. Schneirla, "The Concept of Development in Comparative Psychology," in D. B. Harris (ed.), *The Concept of Development* (Minneapolis: University of Minnesota Press, 1957).

27. Ibid., p. 34. A better analogy might be that of a cake recipe, though as the writers of John Klama (pseudonym), *Aggression: The Myth of the Beast Within* (New York: John Wiley & Sons, 1988), observe: "Where recipes are instructions for what to do with ingredients, genes are physical objects capable of participating in development only when they are part of a functioning gene-environment system," a system capable, too, of regulating them (p. 49).

28. It is noteworthy that, given this idea, the similarity between the present controversy about innateness and the one that has raged, at least since Plato's time, in philosophic circles, becomes so close that it verges on complete identity. After all, the difference between innate information about the environment and innate ideas about God, Platonic forms, and other eternal verities is more or less incidental to the character of the controversy, the main substance of which is that some theorists contend that certain knowledge is innate, while others deny that this is so.

29. Jean Piaget, *Behavior and Evolution* (New York: Pantheon Books, 1978), p. ix.

30. Jean Piaget, *Biology and Knowledge* (Chicago: University of Chicago Press, 1971), p. 215.

31. Gilbert Ryle, *The Concept of Mind* (London: Hutchinson & Co., 1949).

32. W. Craig, "Appetites and Aversions as Constituents of Instincts," *Biological Bulletin* 34, 1918. W. H. Thorpe suggests (in his *Learning and Instinct*) that C. S. Sherrington introduced the same basic distinction in 1906 or earlier.

33. Lorenz, *Evolution and Modification*, p. 82.

34. Niko Tinbergen, *The Study of Instinct* (Oxford: Oxford University Press, 1951), p. 29.

35. Lorenz, *Evolution and Modification*, p. 106.

36. Warren Weaver, "Information Theory to 1951: A Non-technical Review," *Journal of Speech and Hearing Disorders* 17 (1953): 169–70. My critique of Lorenz's use of the notion of information is fairly limited. For a more radical critique of other biologists' uses see Susan Oyama, *The Ontogeny of Information: Developmental Systems and Evolution* (Cambridge: Cambridge University Press, 1985).

37. Francisco J. Ayala, "The Mechanisms of Evolution," *Scientific American*, September 1978, p. 56.

38. My emphasis on mutation is not meant to minimize the importance of other determinants of variation. Still, as E. Mayr remarks: "Most of the genotypic variation found in a population is due to gene flow and recombination. All of it, however, ultimately originated through mutation." See Ernst Mayr, *Populations, Species, and Evolution* (Cambridge, Mass.: Harvard University Press, 1976), p. 105.

39. Richard C. Lewontin, "Adaptation," *Scientific American*, September 1978, p. 213. In a more recent work—R. C. Lewontin, S. Rose, and L. J. Kamin, *Not in Our Genes: Biology, Ideology, and Human Nature* (New York: Pantheon, 1984)—Lewontin rejects this view as an interactionist model that ignores the interpenetration of organism and environment.

40. Weaver, "Information Theory," p. 170. Darwin's theory of "pangenesis" would afford the requisite channels; but there is no empirical support for his so-called gemules, which are supposed to be thrown off by cells and "collected from all parts of the system to constitute the sexual elements." Cited from P. H. Klopfer, *An Introduction to Animal Behavior*, 2d ed. (Englewood Cliffs, N.J.: Prentice-Hall, 1974), p. 12.

41. Lorenz, *Evolution and Modification*, p. 19.

42. Ibid., p. 7. The source, whether phylogenetic or ontogenetic, of the information is presumed to establish whether it is innate or learned. Thorpe, as Hinde remarks (*Animal Behaviour*, p. 317), "emphasizes this point even more strongly, suggesting that it should be possible to determine quantitatively how much of the information . . . comes from the environment and how much was genetically encoded."

43. Ibid., pp. 95–96.

44. Tinbergen, *Study of Instinct*, p. 44.

45. Ibid., p. 45.

46. Ibid., p. 106.

47. Ibid., p. 87.

48. Stanley C. Ratner, "The Comparative Method," in M. R. Denny (ed.), *Comparative Psychology: An Evolutionary Analysis of Animal Behavior* (New York: John Wiley & Sons, 1980), p. 156. From a study by Fabre, excerpted by H. G. Wells.

49. Thomas E. Hagaman, "Insect Behavior: Using the Cricket as a Comparative Baseline," in Denny (ed.), *Comparative Psychology*, p. 231.

50. Lorenz, *Evolution and Modification of Behavior*, p. 12.

51. Ibid., p. 45.

52. Ibid., p. 47. Of course any learning would seem to introduce some plasticity; and W. H. Thorpe argues that "Since comparison involves learning, an element of learning enters into all orientation and all perception." This leads him to suggest "that the difference between inborn and acquired behavior is . . . a difference chiefly of degree of rigidity and plasticity" (*Learning and Instinct*, p. 133).

53. Ibid., pp. 47–48.

Chapter 2

1. Konrad Lorenz, *Evolution and Modification of Behavior* (Chicago: University of Chicago Press, 1965), p. 104.

2. This example was adapted from Daniel Lehrman's own hypothetical illustrations of related problems for Lorenz's position. Cf. his "Semantic and Conceptual Issues in the Nature-Nurture Problem," in L. R. Aronson et al. (eds.), *Development and Evolution of Behavior: Essays in Memory of T. C. Schneirla* (San Francisco: W. H. Freeman and Co., 1970), pp. 28–29.

3. Lehrman, "Semantic and Conceptual Issues," p. 36.

4. Kurt Koffka, *The Growth of the Mind*, 2d ed., trans. R. M. Ogden (London: Routledge and Kegan Paul, 1928), p. 121.

5. Ibid.

6. Koffka, *Growth of the Mind*, pp. 121–22.

7. James Drever, *A Dictionary of Psychology* (Baltimore: Penguin Books, 1952), p. 137. For specific criticisms of a number of concepts of drives see R. A. Hinde, *Animal Behaviour* (New York: McGraw-Hill, 1970). P. H. Klopfer, in *An Introduction to Animal Behavior*, 2d ed. (Englewood Cliffs, N.J.: Prentice-Hall, 1974), suggests that the study of motivation should move away from vague drive theories toward fine-grained behavioral-response analyses that promise "to yield more accurate and testable hypotheses about central processes" (p. 239).

8. Kurt Lewin, "On the Structure of the Mind," in R. W. Marks (ed.), *Great Ideas in Psychology* (New York: Bantam Books, 1966), p. 255.

9. Koffka, *Growth of the Mind*, p. 106.

10. John Dewey, "The Reflex Arc Concept in Psychology," reprinted in *The Early Works of John Dewey, 1882–1898* (Carbondale, Ill.: Southern Illinois University Press, 1972).

11. Koffka, *Growth of the Mind*, p. 104.

12. Ernst Mayr, *Populations, Species, and Evolution* (Cambridge, Mass.: Harvard University Press, 1976), p. 1.

13. As cited and translated in Bertrand Russell, *The Philosophy of Leibniz* (London: George Allen & Unwin, 1900), p. 162.

14. Ibid.

15. Niko Tinbergen, *The Study of Instinct* (Oxford: Oxford University Press, 1951), pp. 87–88.

16. Ibid., p. 87.

17. Lorenz, *Evolution and Modification*, p. 9.

18. So eager is he to claim not merely that such mechanisms can function independently, but that many do, that Lorenz bristles at what he wrongly interprets as Hebb's attributing to both Tinbergen and Lorenz himself the view that only one such process exists. All that Hebb says is, on their view "instinct . . . implies a nervous process or mechanism . . . which . . . is different from those processes into which learning enters" (as cited by Lorenz, *Evolution and Modification*, p. 21). Hebb's use of the indefinite article is interpreted by Lorenz to mean "just one."

19. Lorenz, *Evolution and Modification*, p. 18.

20. Ibid., p. 9.

21. Tinbergen, *Study of Instinct*, p. 84.

22. Jakob von Uexküll, *A Stroll Through the Worlds of Animals and Men: A Picture Book of Invisible Worlds*, in Claire H. Schiller (trans. and ed.), *Instinctive Behavior: The Development of a Modern Concept* (New York: International Universities Press, 1957), p. 6.

23. Whether my preliminary hypothesis is correct is an empirical question, but I might mention, in passing, that it may not be so easily answered experimentally. If the pulling movement is only "released" upon contact of the egg with the center of the lower bill, then my hypothesis might seem defective. However, the sensation of resistance from a stationary egg might be detectable by the goose even if the decreased resistance of the

rolling egg is not (or is, typically, just barely) detectable. There is also a problem here in sorting out initial positioning of the egg from the actual pulling of the egg: If the egg cannot be felt until pulling begins, then felt pressure cannot be construed as a releasing stimulus of the pulling response. Moreover, positioning alone might be the stimulus for pulling even if the goose can feel no pressure from the egg at the center of her lower bill: Sideways-balancing movements insufficient to move a stationary egg might serve to ensure proper initial alignment, and then the pressure on two sides (or even one) might release the pulling. Also, the sight of an egg-shaped object outside the nest might release both the positioning and the pulling behavior.

24. Tinbergen, *Study of Instinct*, p. 84.

25. Ibid.

26. Ibid., pp. 84–85.

27. Wilfrid Sellars, "Philosophy and the Scientific Image of Man," in *Science, Perception and Reality* (London: Routledge and Kegan Paul, 1963), p. 18.

28. Ibid., p. 19.

29. W. V. Quine, "Grades of Theoreticity," in L. Foster and J. W. Swanson (eds.), *Theory and Experience* (Amherst: University of Massachusetts Press, 1970). Although I employ Quine's phrase, I depart from his usage, according to which observation sentences—that is, ones "that we learned to use, or could have learned to use, by direct conditioning to socially shared concurrent stimulation" (p. 3)—constitute the lowest grade of theoreticity.

30. Montaigne's *Essays and Selected Writings: A Bilingual Edition*, trans. and ed. D. M. Frame (New York: St. Martin's Press, 1963), p. 223. More recently, J. A. Fodor relies on "the principle that licenses inferences from like effects to like causes," *Representations* (Cambridge, Mass.: MIT Press, 1981), p. 78. Since causes are often plausibly reckoned as sufficient but not necessary for their effects, perhaps this license should be revoked.

31. Allen Newell, J. C. Shaw, and H. A. Simon, "Report on a General Problem-Solving Program," as quoted in H. L. Dreyfus, *What Computers Can't Do*, rev. ed. (New York: Harper & Row, 1979), p. 76. Given the deterministic character of the computers acting in accord with them, the heuristic rules may still be said to guarantee the outcome, whether successful or not.

32. This claim is owed to the writings of the classic American pragmatists (Peirce, James, Mead, and Dewey), though I would hesitate, if only out of my own ignorance, to credit one of them above the others.

33. Bernhard Katz, "How Cells Communicate," *The Living Cell* (San Francisco: W. H. Freeman and Co., 1965), p. 245.

34. Cf. Leslie L. Iverson, "The Chemistry of the Brain," *Scientific American*, September 1979, for a somewhat recent discussion of these chemicals. Iverson does not make any suggestion, one way or another, about the possible implications of brain chemistry for the brain-computer analogy.

35. René Descartes, *Discourse on Method*, in *Descartes: Philosophical Writings*, trans. N. K. Smith (New York: Random House, 1958), p. 119.

36. Notoriously, Descartes held that only human beings, not brutes, were endowed with a faculty of intellect. "Mindless brute" was for Descartes a redundant expression.

37. I am not suggesting, however, that Descartes's argument is sound; I am just main-

taining that a strictly analogous argument is not in the offing on the present hypothesis. So, too, one might argue that the argument "I walk; therefore, I am" is not strictly analogous to Descartes's version of that argument, that his sort of argument is not available to someone who argues from the premise "I walk."

38. My rhetorically exaggerated suggestion that computer simulations of problem solving substitute products for processes will later be merged with the present suggestion about grasping, into the more unified proposal that flexible prehension governs all genuinely intelligent functioning. A certain kind of flexibility is nowadays claimed for "parallel processing"—in which multiple "series of computation are performed simultaneously" (Dreyfus, *What Computers Can't Do*, p. 320). But even if it does serve better than serial processing to simulate the steps of actual problem solving, parallel computation approaches the genuine article only asymptotically: Relative to "explicit programming," certain structures or "schemas" may seem "emergent" (P. Thagard, *Computational Philosophy of Science* [Cambridge, Mass.: MIT Press, 1988], p. 185), and such emergence (of novelty) may be criterial for flexibility; but the deterministic character of the computer-with-programming allows for reading those schemas back into it. Genuine flexibility requires indeterminacy (and more)—as I hope to show, to render plausible, in the sequel.

Chapter 3

1. This is not to suggest that evolutionary theory is not also well confirmed.

2. Kurt Koffka, *The Growth of the Mind*, 2d ed., trans. R. M. Ogden (London: Routledge and Kegan Paul, 1928), p. 100.

3. Ibid., p. 104.

4. Ibid. Koffka may not worry as much as he should, but that is because he takes himself to have a solution ready to hand.

5. Ibid.

6. Ibid.

7. Ibid., p. 107.

8. Ibid., p. 106.

9. Ibid., p. 107.

10. Since this point may be rather obscure, consider the analogy of a skintight glove: It conforms (or corresponds) exactly to the body but not to the whole body.

11. Koffka, *Growth of the Mind*, p. 251.

12. Ibid., p. 244.

13. Wolfgang Köhler, "The Nature of Intelligence," in M. Henle (ed.), *The Selected Papers of Wolfgang Köhler* (New York: Liveright, 1971), p. 169.

14. For convenience, I am using the term *grasping* as a generic term for all the acts of intelligence Köhler mentions.

15. Köhler, "Nature of Intelligence," p. 171.

16. Ibid., pp. 171–72.

17. Ibid., p. 172.

18. Ibid., p. 173.

19. Ibid.

20. Ibid.

21. Ibid., 174.

22. I am assuming that relations among relations among items are second-order relations among items. If this seems to flirt with paradox while alluding to an ad hoc theory of types to prevent the situation from getting out of control, then there are other ways of arguing the point. Restricting attention to (first-order) relations among (say) just two items in the perceptual field, we could still suggest an infinite number of relations between the items. One such relation would be that item A is twice the distance to the left of item B than point C is to the left of item B. That this relation might be said to amount, extensionally, to nothing more than the relation A is to the left of B is not a particularly telling objection: The example is intentionally intensional.

23. Köhler, "Nature of Intelligence," p. 174.

24. Koffka, *Growth of the Mind*, p. 107.

25. Ibid., p. 251.

26. Köhler, "Direction of Processes in Living Systems," in *Selected Papers*, p. 329.

27. Ibid., p. 331. The pertinence of such observations to functionalism is a topic deserving of special discussion, but such a discussion would be too tangential to the concerns of this essay.

28. Ibid., p. 333.

29. Köhler, "The New Psychology and Physics," in *Selected Papers*, p. 238.

30. Ibid., pp. 238–39. I am extrapolating the premise from other remarks and applying it here, despite the fact that Köhler is here speaking only of "actual human experience" as constituting the range of "local events." This "creative" explication yields an argument that I cannot believe Köhler would have wished to disown: Psychical and behavioral events are within this range.

31. Ibid., p. 243.

32. Ibid., p. 247. Despite appearances, this is not an adverbial analysis of attention: The "objective" is that of attention itself, not the objective (say) of some other activity that one undertakes in a certain manner, namely, attentively.

33. Ibid.

34. Koffka, *Growth of the Mind*, pp. 111–12.

35. Köhler, "Nature of Intelligence," p. 177.

36. Köhler, "New Psychology and Physics," p. 248.

37. Ibid., p. 249.

38. Koffka, *Growth of the Mind*, p. 112.

39. William James, "Habit," *The Principles of Psychology*, vol. 2 (New York: Dover Publications, 1950), p. 109.

40. William James, *Talks to Teachers on Psychology: And to Students on Some of Life's Ideals* (New York: Norton Library, 1958), 57.

41. Köhler, "New Psychology and Physics," p. 248.

42. John Dewey, *Experience and Nature* (New York: Dover Publications, 1958), p. 282.

43. Ibid., p. 295.

44. John Dewey, *Experience and Education* (New York: Macmillan, 1938), p. 35.

45. Ibid. Dewey speaks of another principle, one of *interaction,* as that which "assigns equal rights to both factors in experience—objective and internal conditions" (p. 42). He

conceives of the two principles, continuity and interaction, as "so to speak, the longitudinal and lateral aspects of experience" (p. 44).

46. John Dewey, *Human Nature and Conduct* (New York: Modern Library, 1957), p. 25.

47. Ibid., p. 26.

48. Ibid.

49. Ibid., p. 30.

50. Ibid., p. 33. Note the sameness in difference between Dewey and Köhler on ideas. Both of the authors agree that ideas cannot do what is expected of them without a kind of directional configuration, whether supplied by habit (Dewey) or by native dynamic tendencies (Köhler).

51. Ibid., p. 34.

52. Ibid., p. 37.

53. Ibid., p. 32.

54. Dewey, *Human Nature and Conduct*, pp. 36–37.

55. Dewey, *Experience and Education*, p. 44.

56. Ibid., p. 42.

57. Dewey, *Human Nature and Conduct*, p. 99.

58. Ibid., p. 141.

59. Ibid., pp. 141–42.

60. Ibid., p. 102.

61. These terms, which are usually reserved for feathered bipeds, afford a conveniently brief way to get across some impulsive characteristics of behavior that are claimed by Dewey for humanity.

62. Dewey, *Human Nature and Conduct*, pp. 85–86.

63. Ibid., p. 86.

64. Dewey, *Experience and Nature*, p. 258.

65. Some such needs, including one for oxygen, are met, given what might be, but is not, called the instinctive activity of breathing.

66. Ibid., p. 257.

67. Ibid.

68. Ibid., p. 258.

69. Ibid., p. 272.

70. John Dewey, *The Study of Ethics: A Syllabus*, vol. 4 of *The Early Works of John Dewey: 1882–1894* (Carbondale, Ill.: Southern Illinois University Press, 1971), pp. 236–37. In this work from 1894, Dewey does say that "so far as we can judge" animals lack this capacity of mediation, that their impulses are already coordinated with vital ends (and, hence, are instincts); but he does not explicitly tie their incapacity to the animals' lack of language. My incomplete perusal of this and other early works (e.g., *Psychology: Early Works*, vol. 2) tends to confirm the suspicion that this connection is a later development in Dewey's thinking, inspired perhaps by his association with George Herbert Mead.

71. Dewey, *Experience and Nature*, p. 166.

72. This coined word, modeled after such words as "enables," means "makes possible."

73. Dewey, *Experience and Nature*, p. 166.

74. Dewey, *Human Nature and Conduct*, pp. 99–100.

75. Ibid.
76. James, "Habit," p. 393.
77. Ibid.
78. Ibid., pp. 392–93.
79. Ibid., p. 392.

Chapter 4

1. Gilbert Gottlieb, "Conceptions of Prenatal Behavior," in L. R. Aronson et al. (eds.), *Development and Evolution of Behavior: Essays in Memory of T. C. Schneirla* (San Francisco: W. H. Freeman and Co., 1970), p. 129.

2. Ibid., pp. 129–30. Although partly quoted from Gottlieb, my statement here interpolates the holistic-determination idea from Kuo's own writings, where it is made more explicit.

3. Ernst Mayr, *Populations, Species, and Evolution* (Cambridge, Mass.: Harvard University Press, 1976), p. 416. Susan Oyama views this sort of juxtaposition of claims as insufficiently interactionist, since it tends to "assign formative relevance only to the DNA." *The Ontogeny of Information: Developmental Systems and Evolution* (Cambridge: Cambridge University Press, 1985), p. 26.

4. Ibid., p. 5.

5. G. E. Coghill, *Anatomy and the Problem of Behavior* (Cambridge: Cambridge University Press, 1929). Cited by me from Gottlieb, "Conceptions of Prenatal Behavior." See, too, Zing-Yang Kuo's discussion (of Coghill) in *The Dynamics of Behavior Development: An Epigenetic View* (New York: Random House, 1967).

6. Gottlieb, "Conceptions of Prenatal Behavior," p. 121.

7. Ibid., p. 111.

8. Ibid., p. 123.

9. Ibid., p. 125. All of my remarks are based on Gottlieb's account of Holt.

10. Ibid., p. 93.

11. Ibid.

12. Ibid., p. 97.

13. Ibid., p. 94.

14. Gottlieb, "Conceptions of Prenatal Behavior," p. 130.

15. Kuo, *Dynamics of Behavior Development*, p. 94.

16. Ibid., p. 92.

17. Ibid., p. 95.

18. Ibid., p. 101.

19. Ibid., p. 125.

20. Ibid., p. 126.

21. Ibid., p. 127.

22. Ibid.

23. Ibid., pp. 125–26.

24. Ibid., p. 139.

25. Ibid., p. 197.

26. Philosophical readers will notice that I am relying upon a basically Quinean prin-

ciple about the ontic commitments of a theory: What are said to *be* are the values of a theory's variables, assuming that the values are *real* and not merely possible. This might seem to get us nowhere in clarifying the sense of the (thus) interdependent concepts of *real* and *to be*, but it does indicate more fully the sense of saying that Kuo's theory does not posit behavioral potentials. This simply means that the theory does not introduce any variable specifically limited to having those potentials as its sole values. Nonphilosophical readers can in good conscience ignore these philosophical niceties, if such they be.

27. Gottlieb, "Conceptions of Prenatal Behavior," p. 130.

28. Kuo, "The Need for Coordinated Efforts in Developmental Studies," in *Development and Evolution of Behavior*, p. 181.

29. Kuo, *Dynamics of Behavior Development*, p. 85.

30. Ibid., pp. 19–20.

31. Ibid., pp. 20–21.

32. Gottlieb, "Conceptions of Prenatal Behavior," p. 130.

33. Kuo, *Dynamics of Behavior Development*, pp. 174–75.

34. Karl R. Popper, *The Poverty of Historicism* (New York: Harper & Row, 1964), p. 3.

35. Ibid., pp. 128–29.

36. Kuo, *Dynamics of Behavior Development*, p. 23.

37. Kuo, "Need for Coordinated Efforts," p. 189.

38. Kuo, *Dynamics of Behavior Development*, p. 139.

39. Kuo, *Dynamics of Behavior Development*, p. 11.

40. Kuo, "Need for Coordinated Efforts," p. 189.

41. Kuo, *Dynamics of Behavior Development*, pp. 138–39.

42. Ibid., p. 115.

43. Ibid., p. 131.

44. Ibid., p. 176.

45. Ibid., p. 177. All subsequent statements and illustrative examples of effects are from pp. 177–79 of this text.

46. Ibid., p. 179.

47. Ibid., p. 126. This range is akin to what is sometimes called the "norm of reaction of a genotype"—that is, "the list of phenotypes that will result when the genotype develops in different alternative environments." (R. C. Lewontin, S. Rose, and L. J. Kamin, *Not in Our Genes: Biology, Ideology, and Human Nature* (New York: Pantheon, 1984), p. 268.

48. Ibid., p. 81.

49. Gottlieb, "Conceptions of Prenatal Behavior," p. 130.

50. Ibid., pp. 183–84.

51. Konrad Lorenz, *Evolution and Modification of Behavior* (Chicago: University of Chicago Press, 1965), pp. 23–24.

52. Kuo, *Dynamics of Behavior Development*, p. 199.

53. Daniel Lehrman, "Semantic and Conceptual Issues in the Nature-Nurture Problem," in L. R. Aronson et al. (eds.), *Development and Evolution of Behavior: Essays in Memory of T. C. Schneirla* (San Francisco: W. H. Freeman and Co., 1970), p. 43.

54. Ibid.

55. Kuo, *Dynamics of Behavior Development*, p. 195.

56. Ibid., p. 194.

57. Ibid., p. 188.

58. Ibid., p. 195.

59. Ibid., p. 148.

60. Ibid., p. 81.

61. Ibid.

62. Lehrman, "Semantic and Conceptual Issues," p. 38. Cf. D. D. Jensen, "Operationalism and the question 'Is this behavior learned or innate?'" *Behavior* 17 (1950): 1–8.

63. Lorenz, *Evolution and Modification*, p. 33.

64. Ibid., p. 32.

65. Ibid., p. 33.

66. Lehrman, "Semantic and Conceptual Issues," p. 46.

67. Kuo, "Need for Coordinated Efforts," p. 189.

68. Kuo, *Dynamics of Behavior Development*, p. 26.

69. Ibid., p. 81.

70. Ibid., pp. 200–201.

71. Ibid., p. 201.

72. Ibid., p. 199.

Chapter 5

1. Zing-Yang Kuo, *The Dynamics of Behavior Development: An Epigenetic View* (New York: Random House, 1967), pp. 161–62.

2. Ibid., p. 9.

3. Ibid., pp. 11–12. There is a close parallel here to what Lewontin, Rose, and Kamin propose as a more general biological framework: "The relation between organism and environment is not simply one of interaction of internal and external factors, but of a dialectical development of organism and milieu in response to each other" (*Not in Our Genes: Biology, Ideology, and Human Nature* [New York: Pantheon, 1984], p. 275). This outlook, the authors go on to suggest, is still deterministic overall.

4. R. B. Braithwaite, "Causal and Teleological Explanation," from his book *Scientific Explanation*, as reprinted in J. V. Canfield (ed.), *Purpose in Nature* (Englewood Cliffs, N.J.: Prentice-Hall, 1966), p. 35. Braithwaite's account, suitably revised, will eventually serve as the basis for a credible reconstruction of *instinct*. Other, more recent accounts of teleology do not serve that purpose nearly so well, regardless of their special philosophical merits. Thus, for example, J. Bennett's nonmechanistic account—in his *Linguistic Behaviour* (Cambridge: Cambridge University Press, 1976)—is at odds with the deterministic bent of behavioral nativism.

5. Braithwaite, "Causal and Teleological Explanation," p. 35.

6. Ibid., p. 37.

7. Ibid., p. 38.

8. Ibid.

9. Ibid.

10. Israel Scheffler, "Thoughts on Teleology," in Canfield (ed.), *Purpose in Nature*, p. 55.

11. Ibid., p. 56.

12. Ibid., pp. 56–57.

13. E. S. Russell, *The Directedness of Organic Activities* (Cambridge: Cambridge University Press, 1945), as cited by Braithwaite, "Causal and Teleological Explanation," p. 45.

14. Braithwaite, "Causal and Teleological Explanation," p. 39.

15. I am proposing what might be called a "system property" variant of Braithwaite's account. Cf. E. Nagel, "Teleology Revisited," *The Journal of Philosophy* 74, no. 5 (May 1977): 271.

16. My preceding account of teleology is not to be understood on the model of intentions. I am inclined to agree with Nagel—as against A. Woodfield, *Teleology* (Cambridge: Cambridge University Press, 1976)—that "the intentional view . . . contributes little [internally] to the clarification of the concept as it is used in biology" ("Teleology Revisited," p. 267).

17. A. Rosenblueth, N. Wiener, and J. Bigelow, "Behavior, Purpose, and Teleology," in Canfield (ed.), *Purpose in Nature*.

18. Richard Taylor, "Purposeful and Non-Purposeful Behavior: A Rejoinder," *Philosophy of Science* 17 (1950): 329. As cited by Scheffler, "Thoughts on Teleology," p. 51.

19. John V. Canfield (ed.), *Purpose in Nature* (Englewood Cliffs, N.J.: Prentice-Hall, 1966), p. 5.

20. Braithwaite, "Teleological Explanation," p. 40.

21. Ibid., p. 45.

22. This distinction corresponds to that between processes controlled by "*genetic* and *somatic* programs" (respectively); but an epigenetic viewpoint would seem to suggest that the programs involved in either case are not "*closed*" ("entirely coded in the DNA of the genotype") but "*open programs* that can incorporate additional information." Cf. E. Mayr, *Toward a New Philosophy of Biology: Observations of an Evolutionist* (Cambridge, Mass.: Harvard University Press, 1988), p. 62.

23. A fuller sketch might depict one such mechanism as a negative-feedback device that readjusts itself to make less extensive future corrections when its past corrections have proved too extravagant.

Chapter 6

1. John Dewey, *Human Nature and Conduct* (New York: Modern Library, 1957), p. 66.

2. Gilbert Ryle, *The Concept of Mind* (New York: Barnes & Noble, 1949), p. 42.

3. Ibid.

4. John Dewey, *Experience and Education* (New York: Macmillan, 1938), p. 35.

5. Ryle, *Concept of Mind*, p. 43.

6. Ibid., pp. 43–44.

7. Ibid., p. 50.

8. Ibid.

9. J. L. Austin, "Intelligent Behavior: A Critical Review of *The Concept of Mind*," reprinted in O. P. Wood and G. Pitcher (eds.), *Ryle: A Collection of Critical Essays* (Garden City, N.Y.: Anchor Books, 1970), p. 49.

10. Ryle, *Concept of Mind*, p. 48.

11. Ibid., pp. 135–36.

12. Ibid., p. 138.

13. Ibid. A comparison with Bergson is in order. He contends that laws of science leave out the "duration" essential to life and to the "immediate data of consciousness." Film serves for him as the metaphor for those laws, slicing time into timeless instants and then recreating the illusion of real time by means of a rapid succession of those slices. Film serves Ryle as an analogy of objective "scientific" observation of behavior, but insofar as an illusion concerns him it is not one presumed to be a by-product of the photographic process. It is more the illusion created by the actors. Cf. Henri Bergson, *Time and Free Will* (New York: Harper and Row, 1960).

14. Ibid., p. 139.

15. Ibid. Ryle may rightly discern a connection between excessive distinctiveness accorded to heeding and a relapse into dualism. However, Ryle is wrong about his own enterprise if he thinks that his problem about dualism is not that of needlessly multiplying mental processes but that of conceding the existence of any. In the last quote Ryle should have expressed worries about the consequences of minding's becoming too distinct from both overt and covert activities that are "minded."

16. Ibid., p. 140.

17. Ibid., p. 141.

18. Alan R. White, *The Philosophy of Mind* (New York: Random House, 1967), p. 62.

19. Ryle, *Concept of Mind*, p. 142.

20. Ibid.

21. Ibid., pp. 142–43.

22. Ibid., p. 147.

23. U. T. Place, "The Concept of Heed," in D. F. Gustafson (ed.), *Essays in Philosophical Psychology* (Garden City, N.Y.: Anchor Books, 1964), p. 220.

24. William James, *Talks to Teachers on Psychology: And to Students on Some of Life's Ideals* (New York: Norton Library, 1958), p. 28.

25. Ibid., p. 29.

26. Ryle, *Concept of Mind*, pp. 139–40.

27. Ibid., p. 145.

28. Ibid., p. 144.

29. Ibid., p. 146.

30. Israel Scheffler, "Thoughts on Teleology," in J. V. Canfield (ed.), *Purpose in Nature* (Englewood Cliffs, N.J.: Prentice-Hall, 1966), p. 53.

31. Ryle, *Concept of Mind*, p. 146.

32. Ibid., p. 144.

33. Ibid., p. 145.

34. Ibid., p. 146.

35. Pace Austin, who contends that *purposive* "is a term of psychological art, and to my mind requires some justification: because all our ordinary terminology, not merely the adjectival termination, certainly suggests that intention is related to our action in a more intimate way than its purpose" (J. L. Austin, "Three Ways of Spilling Ink," in his *Philosophical Papers*, 2d ed. [Oxford: Oxford University Press, 1970], p. 280).

36. Scheffler, "Thoughts on Teleology," p. 52.

37. James, *Talks to Teachers*, p. 58.

Chapter 7

1. Richard J. Herrnstein and Edwin G. Boring (eds.), "Functionalism," in *A Source Book in the History of Psychology* (Cambridge, Mass.: Harvard University Press, 1965), p. 482.

2. William James, *The Principles of Psychology*, vol. 1 (New York: Dover Publications, 1950), p. 142.

3. Ibid., p. 141.

4. Ibid., p. 144.

5. Ned Block, "Introduction: What Is Functionalism?" in N. Block (ed.), *Readings in Philosophy of Psychology* (Cambridge, Mass.: Harvard University Press, 1980), vol. 1, p. 172.

6. Herrnstein and Boring, *Source Book in Psychology*, p. 482.

7. Block, "What Is Functionalism?" p. 172.

8. Ibid., pp. 175–76. Block quotes David Lewis, 1966, p. 166.

9. Ibid., p. 173.

10. Ibid., p. 182.

11. Hilary Putnam, "Philosophy and Our Mental Life," in Block (ed.), *Readings in Philosophy of Psychology*, vol. 1, p. 134.

12. This use of one of Putnam's examples is not perhaps for a purpose he would approve of: His example was an analogy; my use of it may be too literal.

13. Gilbert Ryle, *The Concept of Mind* (New York: Barnes & Noble, 1949), p. 47.

14. Ibid., p. 40.

15. Ibid., p. 327.

16. Block, "What Is Functionalism?" p. 176.

17. William James, *Talks to Teachers on Psychology: And to Students on Some of Life's Ideals* (New York: Norton Library, 1958), p. 57.

18. Url Lanham, *Origins of Modern Biology* (New York: Columbia University Press, 1968), p. 217.

19. Putnam, "Philosophy and Our Mental Life," p. 136.

20. A. M. Turing, "Computing Machinery and Intelligence," in A. R. Anderson (ed.), *Minds and Machines* (Englewood Cliffs, N.J.: Prentice-Hall, 1964), p. 17.

21. Ibid., p. 18.

22. Ned Block and Jerry A. Fodor, "What Psychological States Are Not," in Block (ed.), *Philosophy of Psychology*, vol. 1, p. 246. Fodor mentions much the same point in an earlier work—*Psychological Explanation: An Introduction to the Philosophy of Psychology* (New York: Random House, 1968), p. 143—but there he seems prepared to accept the point as a legitimate possibility rather than a fairly damaging objection to functionalism.

23. Ryle expressed such ideas in many of his writings, including ones subsequent to *The Concept of Mind*, on the topic of thinking. Cf., e.g., his essay "A Puzzling Element in the Notion of Thinking," in P. F. Strawson (ed.), *Studies in the Philosophy of Thought and Action* (Oxford: Oxford University Press, 1968). Wittgenstein is often associated with such views and so sometimes regarded, even by some functionalists, as at least a forerunner of contemporary functionalism. Cf., e.g., Stephen N. Thomas, *The Formal Mechanisms of Mind* (Ithaca, N.Y.: Cornell University Press, 1978). In his acknowledgments Thomas says, "Much also is owed to Wittgenstein, whose philosophical instincts sensed the true

locus of difficulties in the philosophy of mind" (p. 5). Is this thought perhaps that, since Wittgenstein has not couched his remarks in terms of a (presumptive) scientific theory T, mere instinct must have dictated them? It's hard to tell.

Chapter 8

1. J. L. Austin, "Three Ways of Spilling Ink," in his *Philosophical Papers*, ed. O. Urmson and G. J. Warnock (Oxford: Oxford University Press, 1970), p. 283.

2. Ned Block, "Psychologism and Behaviorism," *Philosophical Review* 90 (1981): 43.

3. Ibid., p. 20.

4. Ibid., p. 43.

5. Ibid., p. 28.

6. John Dewey, *Human Nature and Conduct* (New York: Modern Library, 1957), p. 66.

7. It is noteworthy, too, that some AI theorists think of parallel computation as giving rise to flexibility when the parallelism is used to simulate complex deliberation. See P. Thagard, *Computational Philosophy of Science* (Cambridge, Mass.: MIT Press, 1988), p. 185. As I understand it, the thought is that many different possible partial solutions are separately, simultaneously generated, then combined and selected. The presumed novelty, which grounds the claimed flexibility, is that of recombination; so the argument I go on to give in the text fits the case of parallel processing.

8. Austin, "Three Ways of Spilling Ink," p. 284.

9. John Dewey, *Experience and Nature* (New York: Dover Publications, 1958), p. 282.

10. Ibid., p. 299.

11. Without attempting a thumbnail sketch of the methods of science and philosophy, I might mention what I here take to be their relevant difference: The former should involve some at least partly observational substantiation of post- or predictions bearing on the validity of some posits and/or claims. Further indication of the form of a philosophical definition is possible—e.g., *one* form it might take is the specification of necessary and sufficient conditions for the applicability of a concept.

12. Austin, "Three Ways of Spilling Ink," p. 285.

13. Ibid., pp. 284–85.

Chapter 9

1. John Dewey, *Experience and Nature* (New York: Dover Publications, 1958), p. 159.

2. Ibid., p. 167.

3. Ibid., p. 168.

4. Jerry A. Fodor, *The Modularity of Mind* (Cambridge, Mass.: MIT Press, 1986), passim. Fodor chooses, on philosophic grounds, not to proffer a definition of modularity, but he does roughly characterize "modular cognitive systems" as "['to some interesting

extent'] domain specific, innately specified, hardwired, autonomous, and not assembled" (p. 37). This (prospectively) theory-laden notion is refined and modified in the course of his book. The unmodified version Fodor attributes to F. Gall: "Modular systems are, by definition, *special purpose* computational mechanisms. . . . [Each of them is a] 'mental organ' . . . pretuned to the solution of computational problems with a specific sort of structure" (p. 120). In my remarks on this notion, I intentionally (in anticipation of my own later proposal that problems grappled with are incomplete teleological schemes) conflate "structure of a problem" and "structure of a module."

5. Ibid., p. 120. Fodor says "that a psychological theory represents the mind as *epistemically bounded* if it is a consequence of the theory that our cognitive organization imposes epistemically significant constraints on the beliefs that we can entertain." My remarks are intended not so much to join issues with Fodor as to exploit his terminology to characterize further conscious readiness—as *non*modular and not definitely epistemically bounded. Fodor concedes the possibility that "someone will some day make serious sense of an unboundedness thesis" (p. 125). I am trying to do so by appealing to the notion of flexibility, but more is needed to make sense of that notion.

6. Gottlob Frege, "Sources of Knowledge of Mathematics and the Mathematical Natural Sciences," in *Posthumous Writings*, ed. H. Hermes et al., trans. P. Long and R. White (Chicago: University of Chicago Press, 1979), p. 267.

7. Gilbert Ryle, *Concept of Mind* (London: Hutchinson & Co., 1949), pp. 149–53.

8. F. N. Sibley, "Seeing, Scrutinizing, and Seeing," in G. J. Warnock (ed.), *The Philosophy of Perception* (Oxford: Oxford University Press, 1967), pp. 139–42.

9. G. W. Leibniz, *Discourse on Metaphysics*, trans. G. Montgomery (LaSalle, Ill.: Open Court Publishing Co., 1962), p. 44.

10. G. W. Leibniz, *New Essays on the Understanding*, in P. P. Wiener (ed.), *Leibniz: Selections* (New York: Charles Scribner's Sons, 1951), p. 402.

11. Ibid., p. 403.

12. This rendering of Chomsky's views is certainly not uncommon: D. W. Hamlyn— in *Experience and the Growth of Understanding* (London: Routledge and Kegan Paul, 1978)—writes: "Chomsky supposes that what the child learns in experience is a set of transformational rules by means of which the basic knowledge of deep structure is mapped on to the surface structure of the actual language he is learning" (p. 32). Still, such a rendering is at best an allowable distortion: Chomsky's "deep structures" are supposed not to be original, unlearned schematisms, but to be the outcomes of prior generative rules that are themselves a product of the encounter, during ontogenesis, between innate schemata and samples of natural language use. What I am calling the "deepest deep structure" is the (allegedly) innate schematism; but it is, in a Chomskian sense, deep and is a structure. Chomsky—in his debate response to Piaget, "The Linguistic Approach," in M. Piatelli-Palmarini (ed.), *Language and Learning* (Cambridge, Mass.: Harvard University Press, 1980)—speaks of such a schematism (in more general terms) as an initial state S_0, "prior to experience, fixed for the species" (p. 109). We may, he says, "think . . . of the initial state as being in effect a function that maps experience onto the steady state" (ibid.). (The latter state, involving generative and transformational rules, is a person's developed linguistic competence.) A state S_0, which is also a function, may pass as a structure, too.

13. Leibniz, *New Essays*, p. 403.

14. Noam Chomsky, *Language and Mind* (Cnicago: Harcourt Brace Jovanovich, 1972), p. 30.

15. The incoherence of this notion becomes obvious when one asks, e.g., whether the set (of proleptically realized structures) would have an equal or odd number of members.

16. Chomsky, *Language and Mind*, p. 100. I am relying here on Chomsky's remark that "the creation of linguistic expressions that are novel but appropriate is the normal mode of language use."

17. Ibid., p. 75.

18. Of course not everyone who subscribes to the anti-instinctive virtues of what I'm calling "flexibility" means by this term what I do, but Stephen Jay Gould does use it in ways I'm happy to second. He writes: "Most of the behavioral 'traits' that sociobiologists try to explain may never have been subject to natural selection at all—and may therefore exhibit a flexibility that features crucial to survival can never display. Should these complex consequences of structural design even be called 'traits'? Is this tendency to atomize a behavioral repertory into a set of 'things' not another example of the same fallacy of reification that has plagued studies of intelligence throughout our century?" (*The Mismeasure of Man* [New York: W. W. Norton, 1981], p. 333.) Their flexibility, as I understand it, effectively precludes reading these behavioral consequences back into us as innate traits.

Chapter 10

1. C. S. Peirce, *Collected Papers*, ed. C. Hartshorne and P. Weiss (Cambridge, Mass.: Harvard University Press, 1934), 6: 20.

2. *G. W. Leibniz: Textes inedits*, ed. G. Grua (Paris, 1948). Quoted from E. M. Curley, "The Root of Contingency," in H. G. Frankfurt (ed.), *Leibniz: A Collection of Critical Essays* (Garden City, N.Y.: Anchor Books, 1972), p. 96.

3. J. L. Austin, "Ifs and Cans," in *Philosophical Papers*, 2d ed. (Oxford: Oxford University Press, 1970), p. 218n.

4. Ibid.

5. Ralph Waldo Emerson, "Fate," in R. E. Spiller (ed.), *Five Essays on Man and Nature* (Arlington Heights, Ill.: Harlan Davidson, 1954), p. 102.

6. Ibid., p. 120.

7. Ibid.

8. Henri Bergson, *Creative Evolution*, trans. Arthur Mitchell (Westport, Conn.: Greenword Press, 1977), pp. 7–9.

9. A model for recent accounts of psychophysical supervenience is Donald Davidson's discussion of "the view that mental characteristics are in some sense dependent, or supervenient, on physical characteristics. Such supervenience might be taken to mean that there cannot be two events alike in all physical respects but differing in some mental respects, or that an object cannot alter in some mental respects without altering in some physical respects." From "Mental Events," in *Experience and Theory*, ed. L. Foster and J. W. Swanson (Amherst: University of Massachusetts Press, 1970), p. 88.

But Davidson's two alternatives don't quite capture the idea that the physical is a sine qua non of the mental—an idea I take to be an essential feature of the supervenience intuition. This feature is brought to the fore of my account—as conscious states' (absolute) dependence for their existence on physical factors. Such an idea is antidualistic and so, in a sense, monistic but not materialistic—one's monism can be ontologically more neutral than that, even if one espouses absolute physical dependency.

Although I wouldn't go so far as to say that psychophysical supervenience should deny that whatever supervenes is physical, I do believe the view isn't intended to imply that physical properties or events supervene on themselves. Davidson-derived accounts of the concept do tend to imply this, but far be it from me to cast the first stone.

Davidson's explication shifts from the term *events* (first alternative) to the term *object* (second), but the shift seems of little consequence. My account (which conjoins something akin to both alternatives) is phrased in terms of *states, features,* and *factors;* but a state (of consciousness) is presumed eventlike, and (physical) factors or features of reality are presumed to be aspects of events in the passing show.

10. This term is taken from Jaegwon Kim's writings on supervenience. My preceding account might be restated, as follows, in virtually the same form as Kim's general characterization of strong supervenience (in "Supervenience and Supervenient Causation," *Southern Journal of Philosophy* 22, supplement [1984]: 49): Consciousness, C^v (i.e., the set of conscious states), *strongly supervenes* on a family of physical complexes, P^v (a "supervenience base" consisting of highly organized sets of physical factors) just in case (i) necessarily for each x and each conscious state, C (in C^v), if x has C, then there exists a physical complex, P (in P^v), such that x has P, and (ii) *necessarily* if any y has P it has C.

Clause (i) is the *absolute dependence* condition, and clause (ii) is the *unique determination* condition. My subsequent remarks (in the text) about possible worlds conceived by God may be presumed implicit in my account as already given (in the text). Those remarks warrant the above Kim-based restatement in modal terms—which give these conditions their counterfactual force.

11. I am glossing over some conceptual concerns that might pose problems for lesser beings, not for God: When I say that twin creations always behave the same, I assume that God is capable of an "induction" from *all* cases; when I say God conceived duplicates for predictive purposes, I assume that any temporal constraints might be circumvented by God's having had the forethought to conceive duplicate worlds with histories aptly noncontemporaneous (from God's frame of reference) to allow for any predictions God might ever choose to make. A Supreme Being might arguably not need to engage in such experiments, but I'll not doubt that He, She, or It *could.*

12. Bertrand Russell, *The Analysis of Matter* (New York: Dover Publications, 1954), pp. 387–88.

13. Ibid., p. 389.

14. Ibid., p. 391.

15. Ibid.

16. Ibid., p. 390.

17. Ibid., p. 392. Most of the things mentioned as bodily movements are, in fact, not just physical but also social in character.

18. Gilbert Ryle, *Concept of Mind* (London: Hutchinson & Co., 1949), p. 51.

19. Geoffrey Hellman and Frank Thompson, "Physicalism: Ontology, Determination and Reduction," *The Journal of Philosophy* 72 (1975): 555.

20. Russell, *Analysis of Matter*, p. 393.

21. Ibid., pp. 391–92.

22. According to Kim's account of strong supervenience, an account to which mine closely conforms (see note 10 above), necessarily anything that has a supervenience base property also has a particular supervenient property. The term *necessarily* may, he sug-

gests, be variously interpreted: "In some cases (e.g., psychophysical supervenience) the most plausible interpretation might be nomological necessity; in other cases, it might be logical or metaphysical necessity; in still other cases, analytical necessity, and so on. For a general definition, the modality is best left as an unspecified parameter to be fixed for each case of application." ("Supervenience and Supervenient Causation," p. 49.) As I have characterized one type of psychophysical supervenience, the modality has been similarly unspecified.

23. Karl R. Popper (and John C. Eccles), *The Self and Its Brain* (New York: Springer-Verlag, 1977), p. 19ff.

24. Ibid., p. 19.

25. Ibid., p. 20.

26. The discussion here relies on various notions from mathematical analysis as presented in Ralph P. Boas, Jr., *A Primer of Real Functions* (Rahway, N.J.: The Mathematical Association of America, 1960), Carus Mathematical Monographs, no. 13. Since the soon-to-be-defined notions of open and closed sets pertain to sets in metric spaces, I've sought to define a distance function that ensures the applicability of those notions. The attempt to give a causal slant to the mathematical notions of a neighborhood and a boundary point is, of course, calculated to render the overall argument favorably pertinent to Russell's causal skeleton hypothesis.

27. Popper, *Self and Its Brain*, p. 35.

28. Ibid.

29. Ibid.

Chapter 11

1. B. Spinoza, *Ethics*, bk. 3, note (as quoted in G. N. A. Vesey, "Agent and Spectator," *Royal Institute of Philosophy Lectures*, vol. 1, 1966–67 [New York: Macmillan, 1968], pp. 140–41).

2. Spinoza, *Ethics*, trans. R. H. M. Elwes, bk. 3, prop. 2, note (New York: Dover Publications, 1955), p. 135.

3. Bertrand Russell, *The Analysis of Matter* (New York: Dover Publications, 1954), p. 391.

4. Pierre Simon de Laplace, *A Philosophic Essay on Probabilities*, as quoted in C. J. Ducasse, "Determinism, Freedom, and Responsibility," in Sidney Hook (ed.), *Determinism and Freedom in the Age of Modern Science* (New York: Macmillan, 1958).

5. Jaegwon Kim, "Epiphenomenal and Supervenient Causation," *Midwest Studies in Philosophy* 9 (1984).

6. Vesey, "Agent and Spectator," p. 154.

7. Cf. Karen Hanson, *The Self Imagined* (London: Routledge and Kegan Paul, 1986) for an elaboration of this theme.

8. Hilary Putnam, "Philosophy and Our Mental Life," in Block (ed.), *Readings in Philosophy of Psychology* (Cambridge, Mass.: Harvard University Press, 1980), vol. 1.

9. W. V. O. Quine, "Two Dogmas of Empiricism," in *From a Logical Point of View*, 2d ed. (New York: Harper & Row, 1963).

10. G. E. M. Anscombe, *Intention*, 2d ed. (Ithaca, N.Y.: Cornell University Press,

1957), p. 11. See, too, her essay "Under a Description," in her *Collected Philosophical Papers*, vol. 2 (Minneapolis: University of Minnesota Press, 1981), for an account of some of the ways her lesson hasn't been heeded.

11. Paul M. Churchland, who favors reconstitution, suggests that "the conceptual framework of modern physical theory" is so "immensely powerful" that "it must be a dull man indeed whose appetite will not be whet by the possibility of perceiving the world directly in its terms." (*Scientific Realism and the Plasticity of Mind* [Cambridge: Cambridge University Press, 1979], p. 7.) Much of the world would escape the notice of anyone who "perceived" things this way.

12. B. F. Skinner, *About Behaviorism* (New York: Random House, 1976), p. 122.

13. Gilbert Ryle, *Concept of Mind* (London: Hutchinson & Co., 1949), p. 82.

14. John von Neumann, *Theory of Self-Reproducing Automata*, ed. A. W. Burks (Urbana, Ill.: University of Illinois Press, 1966), pp. 55–56.

15. Ibid., p. 80.

16. Ibid., p. 78.

17. Ibid., pp. 78–79.

18. Ibid., p. 79. This is the page reference for all subsequent quotes from von Neumann.

19. L. Wittgenstein, *Zettel*, trans. G. E. M. Anscombe (Berkeley: University of California Press, 1970), section 608, p. 106e.

20. The expressions "kind of neurological chaos" and "certain specific chaotic states" may verge on the oxymoronic; but "chaos," as used by Wittgenstein, simply implies a lack of order (or system) among the assembled neurophysiological and biochemical elements, so the expressions might sensibly convey ideas of more or less specific kinds of assemblages that exhibit disorder.

21. The "beast" might still be interpreted epigenetically, of course; so the victory for nativism might be short-lived.

22. Edward O. Wilson, "Academic Vigilantism and the Political Significance of Sociobiology," in A. L. Caplan (ed.), *The Sociobiology Debate* (New York: Harper & Row, 1978), p. 295.

23. Sociobiology Study Group, "Sociobiology—Another Biological Determinism," in *The Sociobiology Debate*, p. 284.

24. E. O. Wilson, "Man: From Sociobiology to Sociology," in *The Sociobiology Debate*, p. 232. Reprinted from *Sociobiology: The New Synthesis* (Cambridge, Mass.: Harvard University Press, 1975).

25. Ralph Waldo Emerson, "Fate," in R. E. Spiller (ed.), *Five Essays on Man and Nature* (Arlington Heights, Ill.: Harlan Davidson, 1954), p. 105.

26. E. O. Wilson, *Sociobiology: The New Synthesis*, p. 562.

27. Wilson, "Man: From Sociobiology to Sociology," p. 233.

28. William James, *Talks to Teachers on Psychology: And to Students on Some of Life's Ideals* (New York: The Norton Library, 1958), pp. 116–31.

Index